The Message of Ecology

The Message of Ecology

CHARLES J. KREBS

University of British Columbia

HarperCollins*Publishers*

Sponsoring Editor: Claudia Wilson
Project Editor: David Nickol
Cover Design: Wanda Lubelska
Cover Photo: Galapagos Penguin by Courtney Milne/Miller Services Limited, Toronto
Text Art: Vantage Art, Inc.
Production Manager: Kewal K. Sharma
Compositor: ComCom Division of Haddon Craftsmen, Inc.
Printer and Binder: R. R. Donnelley & Sons Company
Cover Printer: New England Book Components

The Message of Ecology

Library of Congress Cataloging in Publication Data

Krebs, Charles J.
The message of ecology.

Includes index.
1. Ecology. I. Title.
QH541.K675 1988 574.5 87-19741
ISBN 0-06-043773-1

92 93 10 9 8 7 6 5

To Adam Watson and Robert Moss

Contents

Preface

Ecology is a fascinating subject. This is a book to introduce you to it and the problems ecologists try to analyze. Above all it is an attempt to present the subject in a direct, simple form without including the detail that is necessary in a more conventional textbook and without burdening the subject with abstruse definitions or voluminous statistics. So do not view this book as a text but as supplemental reading designed for an introductory biology course or for a first course in ecology.

You can appreciate the beauty and the sweep of ecological insights without a great deal of complication because humans live in an ecological world where mosquitoes bite and trees die from acid rain. Some ecological insight ought to be in the repertoire of every educated person. Every day you can read about political decisions that have an ecological impact, positive or negative, and your children will inherit a world in which many ecological options are constrained by our present decisions. This book is dedicated to the proposition that *you* need to know some ecology, whether you are now or will be a lawyer, a bus driver, or a computer operator.

So read on! Each chapter ends with some suggestions for further readings should you want to study a topic in more detail. There is a glossary at the end of the book to define unfamiliar words.

I am grateful to Adam Watson and Robert Moss for their comments on the first draft of this book. Alice Kenney helped me survey the literature and prepare the text. Many unnamed ecologists did the research on which this book is based. Perhaps they should be honored as some of the unsung scientific heros of the twentieth century.

Charles J. Krebs

The Message of Ecology

Introduction

Ecology is the science that deals with the interactions of animals and plants in natural systems. The way humans interact with the animals and plants of their surroundings has been critical for their survival during their existence on earth. Early humans were practical ecologists by necessity. Today we live in a technological society that has so changed our life style that many people feel as though they are removed from direct contact with nature. Many think of "nature" as the subject of television programs or as what they look at through the windows of their air-conditioned automobiles. We can make no greater mistake than to ignore or be ignorant of nature. The daily newspapers provide ample illustrations of problems arising from the collision of technology and nature—chemical spills, crop pests, diseases. History provides examples of human cultures that perished because people did not live in concert with nature. Therefore, the first reason for studying ecology is to learn how human societies can survive.

But there is another good reason for studying ecology. We exist in nature, and like all plants and animals we experience wind, rain, heat, and cold, we need food, we need a place in which to live, and we can thus appreciate the problems of ecology from our firsthand experience. Of all the sciences, ecology is perhaps the closest to our daily lives. Ecology deals with animals and plants, as individuals and in groups, and not with entities that we can never see. Therefore, a second reason for studying ecology is that some knowledge of how plants and animals interact is interesting and will enrich your daily life. You will look at our familiar world in new ways.

The science of ecology is concerned with two basic problems: *distribution* and *abundance* of organisms. Distribution is the problem of *where* an organism lives? Penguins live in the Southern Hemisphere, and most species of penguins live on Antarctica or surrounding islands. The ecologist wishes to know why penguins have such a restricted distribution. Why are there no penguins in north polar regions, for example? The problem of abundance reflects the observation that an organism is more numerous in some areas than in others. We ask *how many* individuals live in different areas, and how these numbers may change from year to year. Most of the details of ecological studies can be reduced to these two fundamental problems of distribution and abundance.

Figure 1 illustrates the two problems. If we record the occurrence of a species in the form of a contour map, as in Figure 1a, we can visualize the problems of distribution and abundance with this map. Figure 1b shows the distribution and abundance of pheasants in Illinois. The ring-necked pheasant *(Phasianus colchicus)* is a large game bird that was introduced to North America from Asia in the 1880s. Pheasants do not occur in southern Illinois. The ecologist and the wildlife manager both ask *why?* Pheasants are particu-

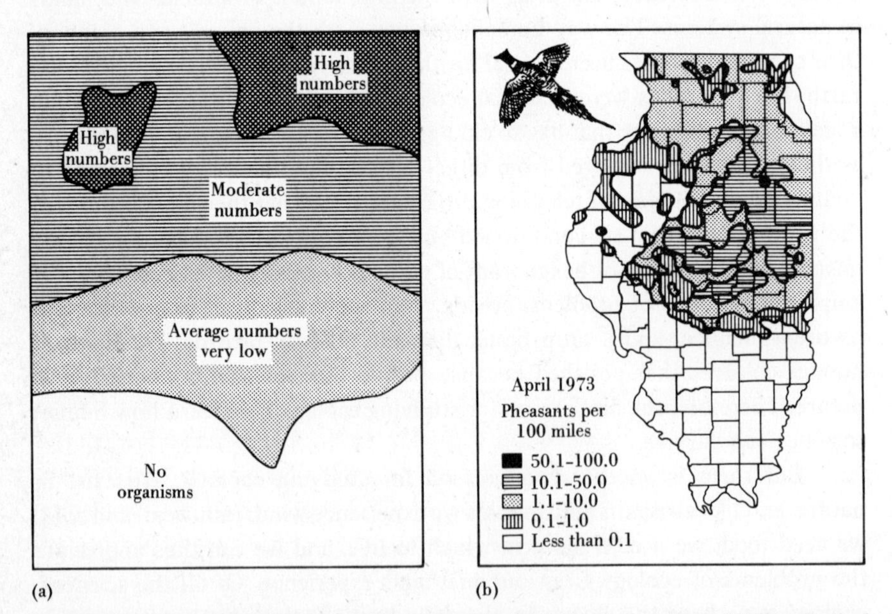

Figure 1 The two central problems of ecology: *distribution* and *abundance.* (a) Hypothetical contour map of the numbers of a species as a general illustration of the two problems. (b) Distribution and abundance of pheasants in Illinois, April 1973. The ecologist wants to know why there are no pheasants in southern Illinois, and why some areas of north-central Illinois support much higher numbers of pheasants than other areas. (After Labisky 1975)

larly abundant in north-central Illinois. The ecologist and the wildlife manager wish to know what makes this area so favorable for pheasants and why some parts of the area seem to be more favorable than others. Note that Figure 1 can be considered a single frame in a motion picture, and we may also ask whether this pattern is stable from year to year. Are the counties that were good for pheasants in 1973 still good in 1986? Changes in the numbers of animals and plants over a period of time are an important component of the problem of abundance.

There are three natural units that ecologists study: *individuals, populations,* and *communities.* A population is all the members of a single species in an area. To explain the distribution and abundance of a population, an ecologist must study what happens to the *individuals* of that species. At this level much specific information on the physiology and behavior of organisms is required. If acid rain is killing spruce trees, what is the mechanism of damage? Are the roots being affected by changes in the soil? Or are the leaves being poisoned directly? The general principle is always the same: If you wish to understand *population* changes, you must analyze what happens to the *individuals* making up the population. I will illustrate how ecologists do this in the first four chapters.

In nature many populations exist together in biological *communities.* An oak forest contains many populations of shrubs, herbs, and fungi, as well as a suite of birds, mammals, and reptiles. Communities are not constants in nature but change because of interactions among the populations and because of disturbances caused by climatic and geological events and by human activity. The explanation of community changes must be done by analyzing how the component populations have altered. I will show examples of community changes in Chapters 5 to 7.

Biological communities are part of the world of physics and chemistry because all organisms are composed of chemicals and obey the laws of physics. We can, however, look at the earth as divided into the physical or nonliving and the biological or living. There are two critical areas of interaction between the physical world and the biological world. One is in the *cycling and recycling of energy and essential nutrients.* All living things need inorganic elements to grow. Where do they get them? How do these elements circulate through the physical and biological worlds? What happens if human activity provides an extra supply to plants? Why do lakes turn green when we dump sewage into them? Biological communities are like giant energy transformers. Plants convert sunlight into complex chemicals through photosynthesis, and the energy in these chemicals becomes food for vegetarians in the animal community. Energy moves on from the plant-eaters into the meat-eaters, and the lion eating a zebra is feeding indirectly on solar energy. I discuss nutrient cycles and how energy flows through communities in Chapter 8. The second critical area of interaction is *climate.* Temperature, moisture, and sunlight control plant growth, and consequently climatic change causes biological change. The ice ages are a most spectacular example, and I discuss the response of communities to climatic changes in Chapter 9.

All organisms have an evolutionary history and the constraints of evolution affect every biological community. Humans did not evolve in polar climates but in tropical heat, and this is why you must wear proper clothing if you wish to go skiing. Tropical plants cannot live with frost, and the seeds of some temperate zone plants cannot germinate without frost. In the last chapter I discuss how evolutionary history can help explain the present-day distribution and abundance of animals and plants.

Our knowledge about distribution and abundance has expanded greatly in the last 50 years as ecologists have analyzed in detail the mechanisms that affect the numbers of organisms. In the next 10 chapters I attempt to summarize in 10 ecological precepts the main results of all this scientific effort. I have at times loosely referred to these precepts as the Ten Commandments of Ecology, but they are probably better thought of as ecological generalizations that tell us how natural systems operate. They form the message of ecology.

chapter 1

The Distribution of Species Is Limited by Barriers and Unfavorable Environments

Polar bears occur neither in New York nor in Antarctica. We are not particularly surprised about their absence in New York. Polar bears hunt off the north polar ice pack for seals, and New York is too far south for the polar ice to reach. But polar bears live happily in the New York zoo, so clearly the climate of New York is not the restricting factor. More likely it is the food supply of seals that a polar bear can catch. We should be more surprised that polar bears do not live in Antarctica, since it abounds with both ice packs and seals, yet the reason is simple. Polar bears have never reached the antarctic region because the tropical oceans form a *barrier* that they cannot cross.

Dispersal means movement, in particular the movement of an individual from its place of birth to a new place for breeding and reproduction. Movement is crucial in many ecological situations, but nowhere are the effects of movements more clearly shown than in the study of distribution.

In the 1830s Charles Darwin traveled the globe in the HMS Beagle to survey the variety of plants and animals unique to places like South America and the Galapagos Islands. Darwin possessed the insight to see that evolution in isolated populations had produced these great differences in flora and fauna. Isolation, or lack of dispersal, thus became a cornerstone of the Darwinian view of how the animals and plants of the world came to be. Dispersal, or rather a lack of dispersal, is thus the reason we go to Africa to see giraffes and not to South America, and why we go to Australia to see kangaroos and not to North America. Our zoos are thus a popular monument to the role of dispersal in affecting the distribution of animal life on the globe.

But a problem arises here. Evolution has certainly produced different plants and animals in different geographical realms, but what assurance do we have that any one of these organisms could in fact live in a quite different area? This question can be answered very simply by a *transplant experiment*—move the organism to a new area. If it survives there and reproduces, you have good evidence that the former distribution was restricted by a lack of dispersal. Figure 1.1 illustrates the logic of the simple transplant experiment.

Are there any examples of transplant experiments in real life? People have carried out transplant experiments, often inadvertently, since the earliest times, but in the last two centuries this trickle of transfers has turned into a flood. Many of our serious pests are introduced species, and the ecology of invasions has economic impact on our lives. Many of the species transplanted are accidentals—seeds caught in bales of wool, mice transported in bales of hay. Elaborate inspection and quarantine procedures in different nations illustrate how people strive to prevent the accidental introduction of organisms harmful to humans and their domestic animals.

Paradoxically, some of the worst pest species have been introduced deliberately. Consider just two examples. The European starling *(Sturnus vulgaris)* has spread over the entire United States and much of Canada within a period of 60 years. The starling is considered a pest because it is bold and aggressive, attacks some fruit crops, and has displaced several native bird species. Originally it occurred in Eurasia, from the Mediterranean to Norway

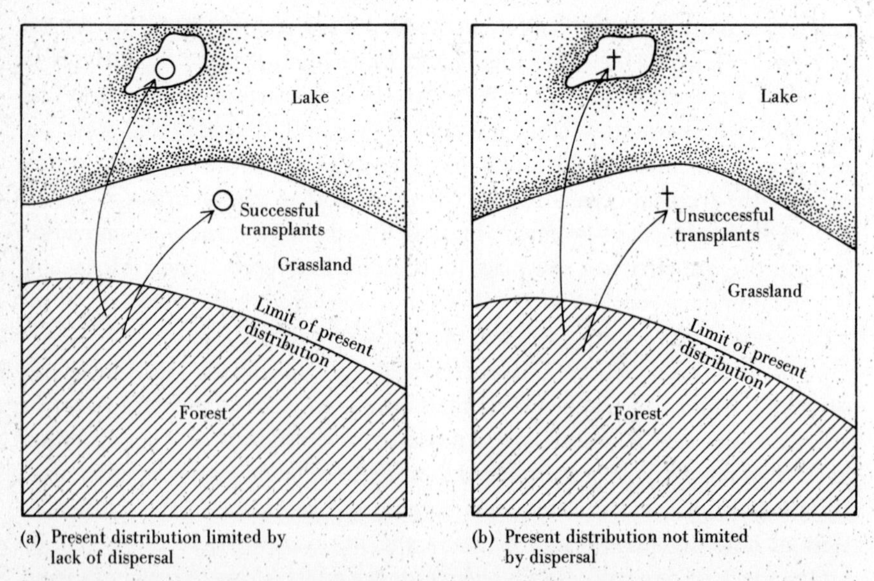

Figure 1.1 Illustration of how transplant experiments can be used to discover whether the present distribution of an animal or plant is restricted by dispersal barriers. In this hypothetical example a species is found only in forest areas and not in grassland.

and east to Siberia. Many attempts were made to introduce the starling into the United States. One attempt was made in West Chester, Pennsylvania, before 1850, another was made in Cincinnati, Ohio, in 1872–1873, and 20 pairs were released in Portland, Oregon, in 1889, but nothing came of these or several other importations. No one knows why these early introductions failed—perhaps too few individuals were released.

The permanent establishment of the starling dates from April 1890, when 80 birds were released in Central Park, New York City. In March of the following year 80 more were released. About 10 years were required for the starling to become established in the New York City area. It has since expanded its range across North America (Figure 1.2). The breeding range expands through the irregular migration and wandering of nonbreeding juvenile birds, 1 and 2 years of age. Adult starlings typically use the same breeding area from year to year and thus do not colonize new areas. About 3 million

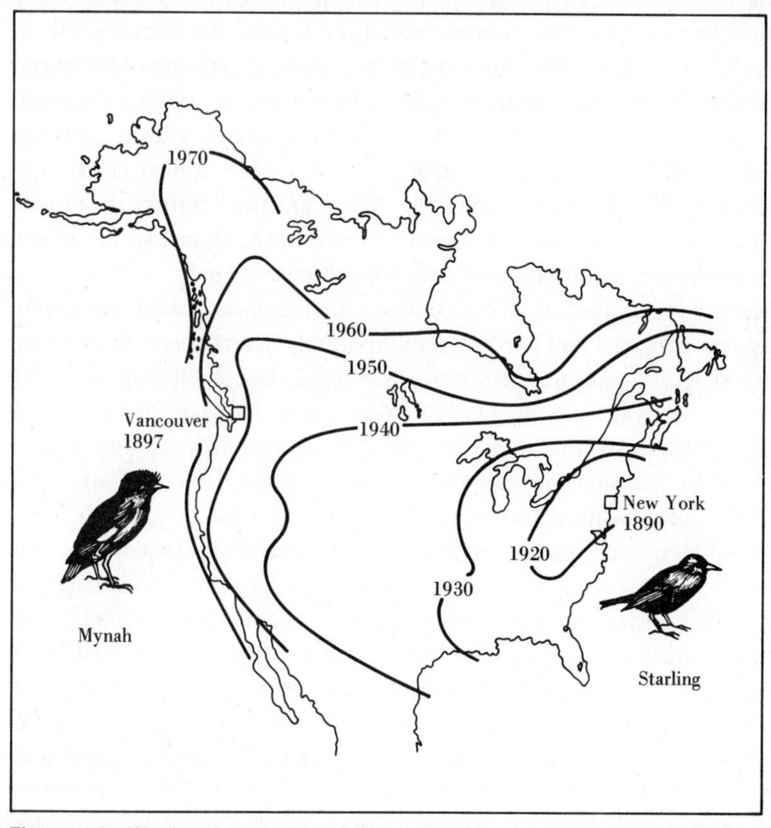

Figure 1.2 Westward expansion of the range of the European starling *(Sturnus vulgaris)* in North America. The starling was introduced to New York in 1890, and the limits of its breeding distribution in subsequent years are shown. The crested mynah *(Sturnus cristatellus)* was introduced to Vancouver about 1895, but has not spread since. (After Johnson and Cowan 1974)

square miles were colonized by the starling during the first 50 years after its successful introduction, and a bird unknown to our forefathers has now become one of the more common birds in North America.

The European rabbit *(Oryctolagus cuniculus)* was originally found in the western Mediterranean area. It began to spread to new areas as early as the eleventh century B.C., when Phoenician traders carried rabbits on their ships as a source of fresh meat. In Roman times it was realized that the rabbit could be a serious pest because of its burrowing and grazing, but traders continued to transport the rabbit throughout the world. The European rabbit now lives in as diverse places as some of the Hawaiian Islands, Macquarie Island in the Antarctic, San Juan Island in Puget Sound, and Australia.

Let us follow the story of how rabbits were introduced into Australia. European rabbits reached Australia with the first European settlers in 1788 and repeated introductions followed. By the early 1800s rabbits were being kept in every large settlement and had been liberated many times. All the early rabbit introductions either died out or remained localized. No one knows why.

On Christmas Day, 1859, the brig HMS *Lightning* arrived at Melbourne with about a dozen wild European rabbits bound for an estate in western Victoria. Within three years rabbits had started to spread, after a bush fire destroyed the fences enclosing one colony. From a slow spread at first the colonization picked up speed during the 1870s, and by 1900 the European rabbit had spread 1000 miles to the north and west, changing the entire economy of nature in southeastern Australia (Figure 1.3).

The rapid colonization of Australia by the European rabbit was partly due to natural dispersal and partly a result of human interference. Spurred by the fur trade of the 1880s, trappers carried rabbits in their saddle bags to found new colonies in advance of the wave of colonization. By 1910 the rabbit had occupied most of the southern half of Australia, an area of 1.5 million square miles, living in environments ranging from arid, stony deserts to subalpine valleys, from wet coastal plains to subtropical grasslands.

The rabbit quickly became the most serious agricultural pest ever known in Australia. Rabbits eat grass, the same grass used by sheep and cattle, and so quickly the cry went up from the pastoralists: "Get rid of the rabbit!" The subsequent history of control attempts in Australia is a sad chronicle of ecological ignorance. Millions of rabbits were poisoned and shot at great expense with virtually no effect on their numbers. Nowhere else has the introduction of an exotic species had such an enormous economic impact and spotlighted the folly of the introduction experiment: *Act in haste, and repent at leisure.*

Not all transplant experiments have harmful results, and one of the challenges of ecology is to sort out the positive and the negative *before* the transplant is done. Many fishes have been introduced into new areas success-

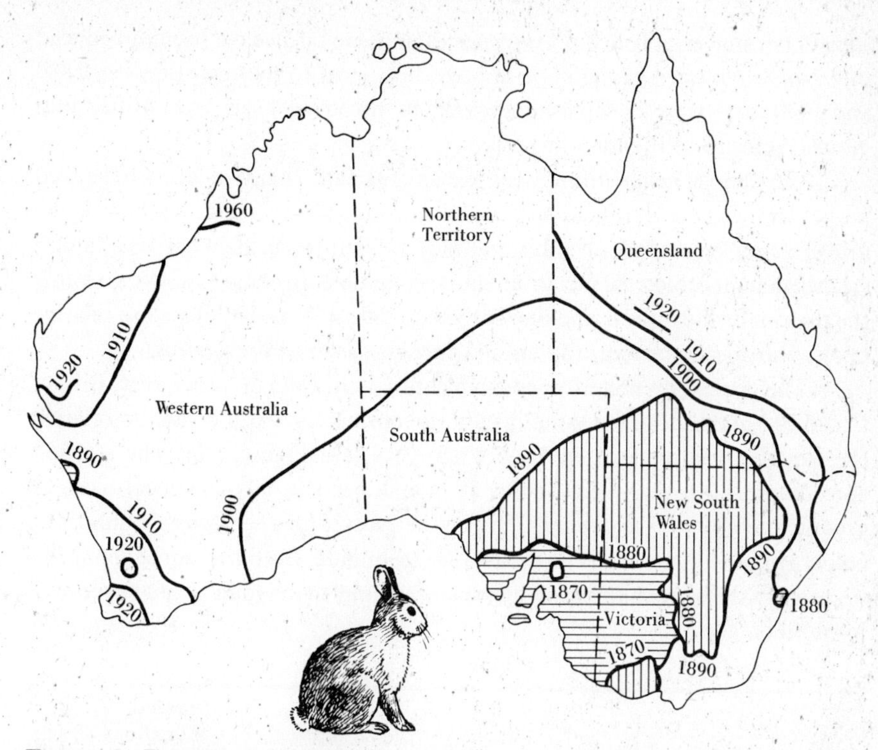

Figure 1.3 The pattern of spread of the European rabbit *(Oryctolagus cuniculus)* in Australia, 1870–1960. (From Myers 1971)

fully, with a resulting improvement in fishing. The rainbow trout, a prize game fish among fishermen, is a native of cool rivers and streams of western North America. Although originally the rainbow trout did not occur east of the Continental Divide in North America, it now occupies streams in all the Canadian provinces and most of the states, as well as some of the river systems in Mexico and Central America. It has been introduced all over the globe during the last hundred years and is now firmly established on all continents except Antarctica. Trout fishing has expanded greatly because of these introductions. But even such an apparently desirable introduction may have undesirable side effects in some regions. For example, rainbow trout can displace native brook trout, another prized game fish, in the southern Appalachians.

Not all transplant experiments are successful, and the dramatic effects of the successful transplants, such as the rabbit in Australia, tend to overshadow the humdrum failures of many other introductions. The European rabbit has been introduced many times into continental United States, but (fortunately!) none of these introductions has been successful. Bird introductions into continental areas are usually failures. In North America only 4 of 50 introduced bird species are common. In Europe only 13 of 85 introduced species became established. Many species of game birds have been introduced into North America in the hope of providing sport to hunters. Between 1883 and 1950 there were 23 separate attempts to introduce 4 species of grouse into North America. All ended in failure.

We almost never know why transplants fail. Many of our introduced pests, such as the starling, had to be introduced several times before the introduction succeeded, and this makes it difficult to decide whether a failure is a permanent ecological fact or an unlucky gamble. In some cases we do know the reason for the failure, and one of the best examples is that of a close relative of the European starling, the crested mynah *(Sturnus cristatellus).*

The crested mynah escaped from captivity in 1895 in Vancouver, British Columbia. At almost the same instant, a continent's width away, its relative the starling was released in New York City (see Figure 1.2). The starling spread across North America, but the mynah persists only as a local resident around the city of Vancouver. Part of the reason for the mynah's failure to colonize North America is its incubation technique. Both the starling and the mynah lay the same number of eggs per nest, but the mynah is less successful in producing young.

	European starling	Crested mynah
Average number of eggs laid per nest	5.2	5.1
Percentage of eggs hatched	82	58
Percentage of young leaving nest	69	38

Mynahs are tropical birds that nest in holes in trees and incubate their eggs only irregularly. This behavior is successful in warm Hong Kong, their native home, but it is not adaptive in cool Vancouver. If mynah eggs are moved to a starling nest, 90 percent hatch. The clinching experiment was the controlled artificial heating of Vancouver nesting places of some mynahs to Hong Kong temperature; 92 percent of the eggs hatched in the heated boxes compared to only 64 percent in the unheated nests (Figure 1.4). Thus the crested mynah has failed to adapt its incubation rhythm to temperate zone conditions, and partly because of this it has not been able to increase its geographical range.

Transplants may fail for two general reasons: either the *biological environment* eliminates the newcomer or the *physical-chemical environment* is lethal to the organism or prevents it from reproducing. *Predation* is an element in the biological environment that affects the establishment of some species. A good illustration of the role of predators can be seen in the common mussel *(Mytilus edulis)* which lives attached to rocks along sea coasts throughout the world. On the exposed southern coast of Ireland small mussels are abundant,

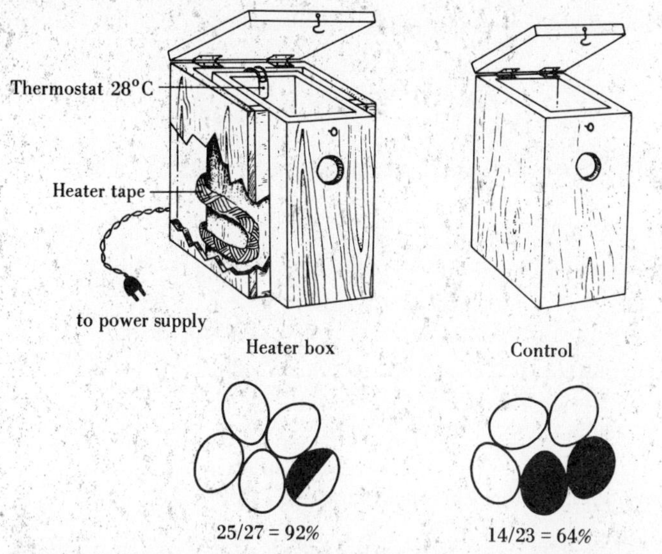

Figure 1.4 Crested mynah experiment with altered nest microclimate in Vancouver. At five crested mynah nests a heater was installed and nest temperature was maintained at Hong Kong levels (28° C). Hatching success at these nests is contrasted with the controls exposed to natural temperature fluctuations. (After Johnson and Cowan 1974)

but in protected waters mussels are often absent. The reason for this can be seen easily if one moves pieces of rock with mussels attached from exposed coast to protected waters (Figure 1.5). Mussels disappear rapidly from protected waters because they are eaten by three species of crabs and a starfish. If you place the transplanted mussels inside wire mesh cages, they live happily,

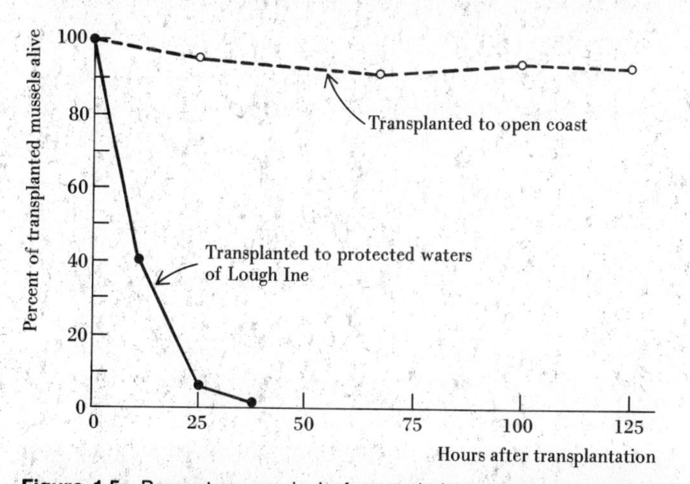

Figure 1.5 Percentage survival of mussels in transplant experiments on the Irish coast. Small mussels disappear rapidly when transplanted anywhere in Lough Ine, but do not disappear if transplanted to the open coast. Crab and starfish predators are able to open and eat mussels in sheltered waters. (After Kitching and Ebling 1967)

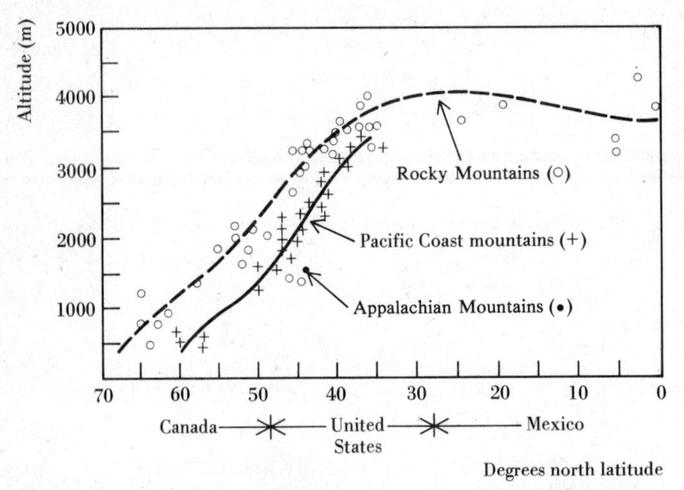

Figure 1.6 Timberlines in North America. (After Daubenmire 1954)

as long as the predators cannot get into the cage. The crabs and the starfish are uncommon on the open coast because of heavy wave action in the intertidal zone, and the mussels thus have a refuge where they are relatively safe.

Many other biological interactions, such as *competition,* can affect the distribution of a species on a local scale. Chemical warfare is used by many organisms to suppress possible competitors. A good example of this is the action of penicillin, the secretion of a fungus, on microorganisms.[1] The study of human disease is essentially a study of colonization (by microorganisms) of new environments (people), and thus differs only in scale from the starling's colonization of North America. At some time in our lives most of us owe a debt to the chemical warfare of an antibiotic against some disease organism, and the restriction and elimination of the invading microbe in our bodies. Many other plants secrete toxic chemicals that inhibit other plants or the animals that try to feed on them, and the study of chemical warfare is a new and active field of research in ecology (see Chapters 3 and 10).

The upper limit of trees on a mountain, often called the *tree line* or *timberline,* is a particularly graphic illustration of a distributional boundary that is controlled by the physical environment. Three major factors change as one proceeds up a mountain—temperature decreases, rainfall increases, and wind velocity increases. Because of freezing temperatures during much of the year, little soil moisture may be available to plants, and soil formation itself may be slow.

In North America timberlines are remarkably structured (Figure 1.6). Timberlines decrease about 110 meters in altitude for each degree of latitude

[1]In nature the soil fungus *Penicillium* excretes this antibiotic to protect itself against bacteria. Humans have simply learned to use this chemical for their own protection.

(approximately 110 kilometers) as one moves north from the United States–Mexico border. Between this border and the equator timberlines are constant around 3500 to 4000 meters. Within North America at any given latitude, timberlines are lowest in the east in the Appalachian Mountains and highest in the Rocky Mountains. The uniformity of timberlines shown in Figure 1.6 is somewhat surprising because many different species of trees are involved.

Trees at upper timberline in the Northern Hemisphere are often windblown and dwarfed (Figure 1.7), and this might suggest wind as a major factor limiting trees on mountains. Wind velocity certainly increases rapidly as one goes up mountains, but there are two difficulties with the wind idea. Within the tropics and in the Southern Hemisphere, wind effects seem to be absent, yet these mountains still show a timberline. A more fundamental objection is that all the evidence about wind is relevant to *old* trees, whereas it is the establishment of the very young seedlings that is crucial to timberline formation. Almost no one has experimentally transplanted seeds or seedlings of trees above timberline to see why they fail to take hold. Adequate soil moisture may be the critical element for small seedlings, and a combination of low temperature, which allows only slow growth, and shortage of soil moisture may condemn most tree seedlings to an early death. During the last 70 years small trees have become established at timberline on Mount Rainier and Mount Baker in the Cascades of western Washington only during a few exceptional years with warmer than average temperatures. Temperature may thus hold the key to the *establishment* of trees at alpine timberlines. The subsequent *survival* of these seedlings is controlled by winter desiccation or frost drought.

The forest-prairie boundary in North America formerly extended from Alberta to Texas in a complex transition (Figure 1.8 on page 16). This boundary, now obscured by agricultural activities, was remarkably abrupt. When settlers began to colonize the Great Plains, they could pass from closed forest to grassland in a few meters. Numerous tongues of forest extended far out into the grassland along river valleys, and isolated patches of prairie extended as far east as Indiana and Ohio. Here, then, is another distributional boundary written plainly across a continent. What prevented deciduous trees from colonizing the prairies?

Two reasons have been suggested why the prairies of North America are treeless: fire and shortage of precipitation. Fire is an important factor restricting trees because fires kill tree seedlings, but do not damage grasses. Trees grow from the top buds, while grasses grow from the base, which is under the ground surface and protected from fire damage. Early explorers and settlers almost without exception commented on the extensive prairie fires. Some of these fires may have been set by Indians, but most were caused by lightning.

The summer climate of the grasslands is characterized by periodic summer droughts. The dust bowls of the 1930s were a highly visible product of

Figure 1.7 Dwarf growth forms of Engelmann spruce at timberline (3000 meters) in the Rocky Mountains of Colorado. (a) Tall conifers of subalpine fir and Engelmann spruce give way to dwarf forms called elfinwood or krummholz as one ascends the ridge (Rogers Pass).

Figure 1.7 (continued) (b) Wind-shaped dwarf form of Engelmann spruce found at 3050 meters on Niwot Ridge, Colorado. (Photos courtesy of J. Mitton)

a typical period of summer prairie drought. Severe prairie droughts have recurred at about 20-year intervals during the last century. Trees survive well in grassland areas only in the wet years and die back in years of drought. During the drought of the 1930s, 50 to 60 percent of the trees in Nebraska and Kansas died. The grassland climate with its summer drought thus favors both fires and grass, and together these physical effects set the prairie-forest boundary. Fire tipped the balance against trees and in favor of grass.

Let me try to draw together some of the threads that I have developed in this chapter. Every animal and plant species has a restricted distribution on the globe, and we would like to know why. The barriers to dispersal—land, water, mountains—set the broad patterns of life on earth, and the resulting diversity of organisms on isolated land masses was a major stimulus for Darwin's theory of evolution. Barriers to dispersal have thus set the broad limits to distribution on a global scale, and the widespread movement of exotic species by humans from one continent to another in the last 200 years has produced a score of pest problems that are permanent monuments to ecological ignorance.

On a more local scale a range of physical and biological constraints operates to prevent organisms from increasing their geographical range. As distances become shorter and barriers less important, dispersal becomes less significant as a limiting factor, and the transplant experiment more often fails.

Some practical advice follows from all this. *No one should be allowed to introduce any exotic animal or plant from one continent to another.* The only exception should be species that can be shown by rigorous research beforehand

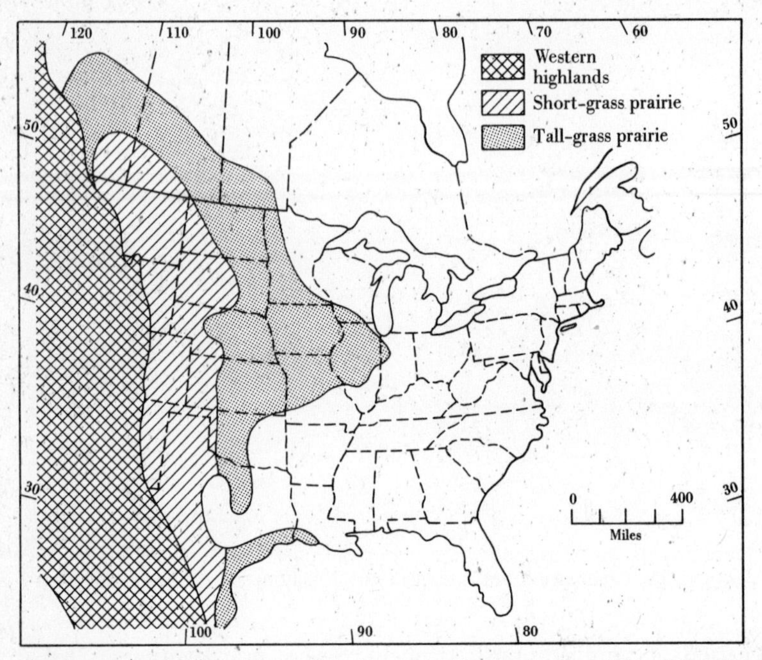

Figure 1.8 Grassland of North America east of the Rocky Mountains. (After Borchert 1950)

not to cause damage to the environment or to other native organisms. The burden of ecological proof must be shifted to the person who suggests the introduction, and we should stop the ecologically naive practice of assuming that introduced species are harmless additions. We should practice the same restraint with all organisms that we apply to the disease organisms that affect us and our domestic animals.

FURTHER READING

Arno, S. F. 1984. *Timberline: Mountain and Arctic Forest Frontiers.* The Mountaineers, Seattle, Wash.

*Elton, C. S. 1958. *The Ecology of Invasions by Animals and Plants.* Methuen, London.

George, W. 1962. *Animal Geography.* Heinemann, London.

King, C. 1984. *Immigrant Killers: Introduced Predators and the Conservation of Birds in New Zealand.* Oxford University Press, Oxford.

Ridley, H. N. 1930. *The Dispersal of Plants Throughout the World.* Reeve, Ashford, Kent, England.

Ritchie, J. 1920. *The Influence of Man on Animal Life in Scotland.* Cambridge University Press, Cambridge.

Wallace, A. R. 1876. *The Geographical Distribution of Animals.* London.

*Highly recommended.

chapter 2

No Population Increases Without Limit

Since everything we know, from our paychecks to our patience, is limited, we are not particularly surprised to find that populations of plants and animals are also limited in numbers. Aristotle pointed this out 2300 years ago, Darwin repeated it 130 years ago, and still we do not understand the ramifications of this simple observation. Right now you may well regard this ecological statement as a trivial truism. The next 10 pages may convince you that it conceals a staggering ecological complexity that we only dimly comprehend even today.

A population is a group of interbreeding organisms belonging to the same species. If populations are limited in numbers, what prevents populations from increasing? To answer this question, we break down population changes into four processes.

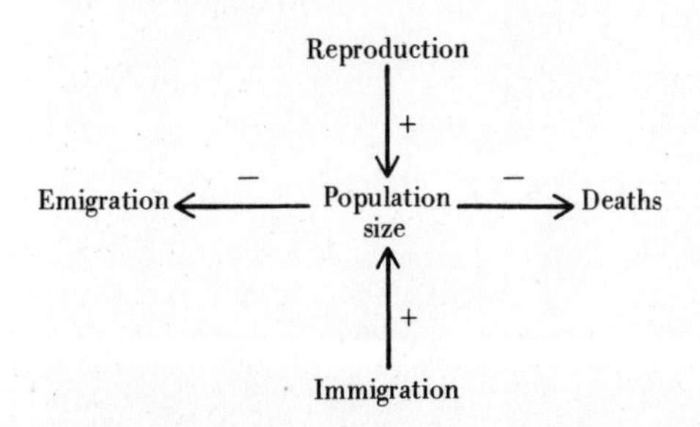

Two processes add animals or plants to a population: *reproduction,* the addition through births or seed production, and *immigration,* the addition through movement into a population. Two processes remove organisms from a population: *deaths* and *emigration.* If we want to find out why a population has increased, we can now reduce this question to a determination of which of the four processes has changed to allow increase. All populations are subject to natural controls that in the long run balance out births, deaths, immigration, and emigration so that population size stays roughly the same. But in the short run, paradoxically, we see little evidence of this balance. Birds that are common one winter may be rare the next. Garden pests that are bothersome one summer are nowhere to be seen the following year. The "balance of nature" seems to disappear when we look closely for it. The nineteenth-century concept of balance was the simple idea that populations in nature are constant. It had to be replaced once we discovered that fluctuations in natural populations are commonplace. In the process ecologists have uncovered a much more interesting model of the balance of nature.

Population analysis starts by sorting out births, deaths, immigration, and emigration, and then finding out what controls these processes. Four components of the environment act to change births, deaths, immigration, and emigration: (1) weather, (2) food, (3) other organisms, and (4) a place in which to live. For any particular population one of these four components may be of overriding importance in preventing population increase. Consider a few examples of population analysis.

Figure 2.1 Swarm of desert locusts *(Schistocerca gregaria)* in Morocco, 1954. (Photograph courtesy of the Food and Agricultural Organization of the United Nations and the Anti-Locust Research Centre, London)

Locusts are pests of long standing, and locust plagues, described in the Old Testament, have been a recurrent disaster for agricultural societies. Locusts are similar to grasshoppers, but at times they swarm and migrate in mass bands, which concentrates their destructiveness (Figure 2.1). A locust eats approximately its own weight in green food each day, and little that is green remains after a swarm passes.

There are few species of locusts in the world, and a list of 8 to 10 covers all the swarming species known. One of the most destructive is the desert locust *(Schistocerca gregaria)* which occupies Africa north of the equator, the Middle East, and India. The biblical plague of locusts in Egypt was probably this species. Since 1908 there have been four major plagues of the desert locust (Figure 2.2). High populations last for 7 to 13 years and then collapse to low numbers for periods of up to 6 years.

It was not until the early 1900s that any serious ecological work was begun on locusts. A Russian entomologist, B. P. Uvarov, discovered in 1913 that locusts can exist in two extreme forms. At one extreme is a pale-colored, short-winged form that lives alone, like a solitary grasshopper. Uvarov called this phase *solitaria.* At the other extreme is a dark-colored, long-winged form that is a gregarious, swarming locust. He called this phase *gregaria* (Figure 2.3). Uvarov discovered that these extreme forms are connected by a series of intermediates and are thus phases of the same species.

The development of a locust plague involves two conditions: *weather* and *phase transformation.* A favorable environment allows the numbers of locusts to increase. Weather, particularly moisture, is the most important aspect of the environment. The main breeding areas of the desert locust in northern Africa have scanty and erratic rainfall, so moisture is often limiting. Moisture is required for egg development; the eggs must be laid in moist soil to survive. Green food is needed by both nymph and adult locusts, and moisture is again critical here. But, in addition to favorable weather, locusts must go through the process of phase transformation before they can swarm, and we thus need to find out how phase *solitaria* locusts are transformed to phase *gregaria* locusts.

Phase transformation is first and foremost a behavioral change. *Solitaria* locusts do not react to the presence of other locusts, while *gregaria* locusts prefer being together. We do not understand how ecological events in field populations lead to concentration and aggregation of locusts. In the laboratory we can artificially crowd locusts together in small cages and produce some of the changes in color and size from *solitaria* to *gregaria.* Aggregation is assisted by a chemical called *locustol,* which is produced in the intestines of immature locusts, passes out in the feces, and evaporates. Locustol in the air causes immature locusts to be more gregarious and to change color toward the *gregaria* phase. The effects of crowding may be cumulative from generation to generation. A complete phase transformation does not seem possible within

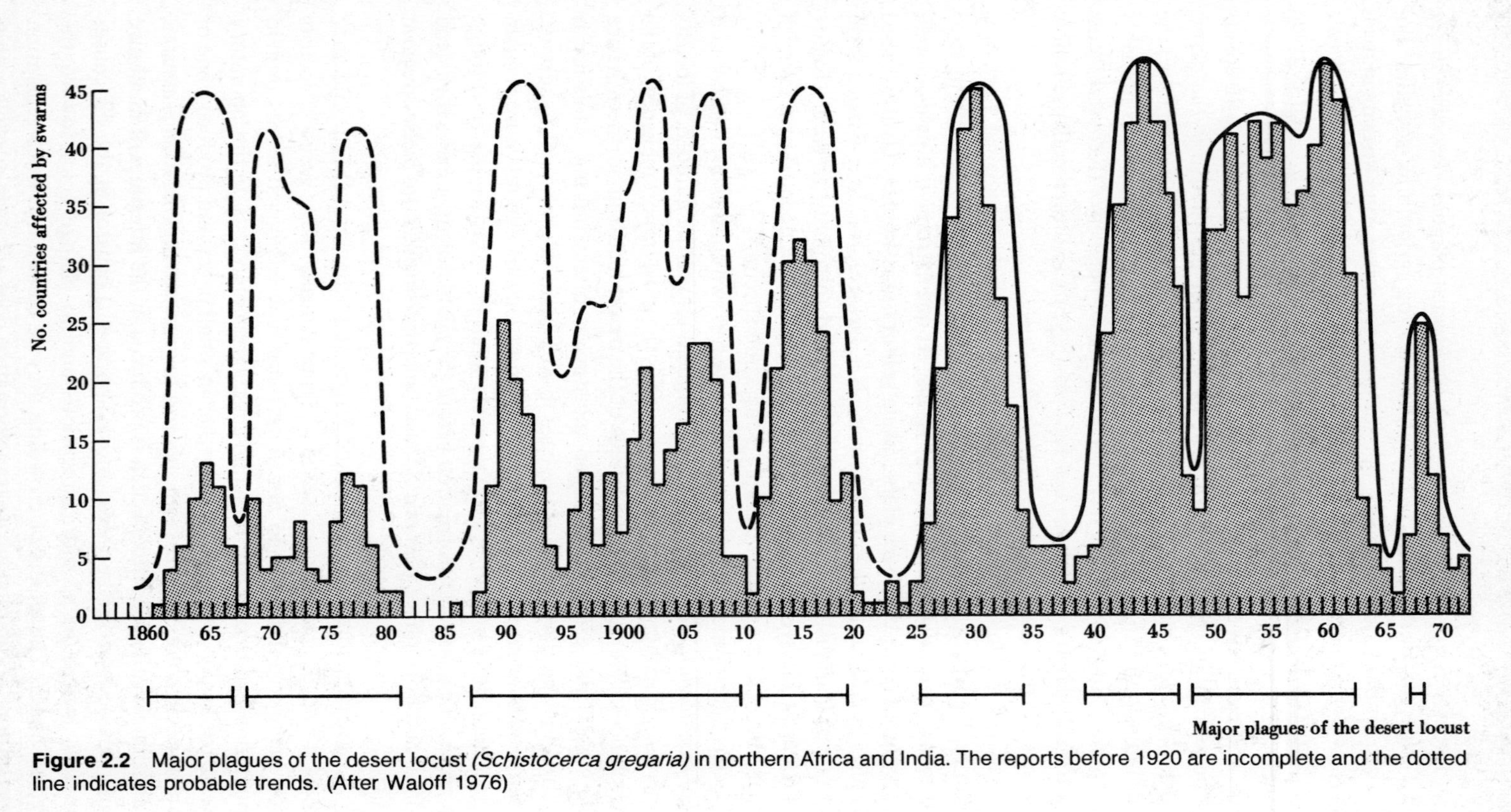

Figure 2.2 Major plagues of the desert locust *(Schistocerca gregaria)* in northern Africa and India. The reports before 1920 are incomplete and the dotted line indicates probable trends. (After Waloff 1976)

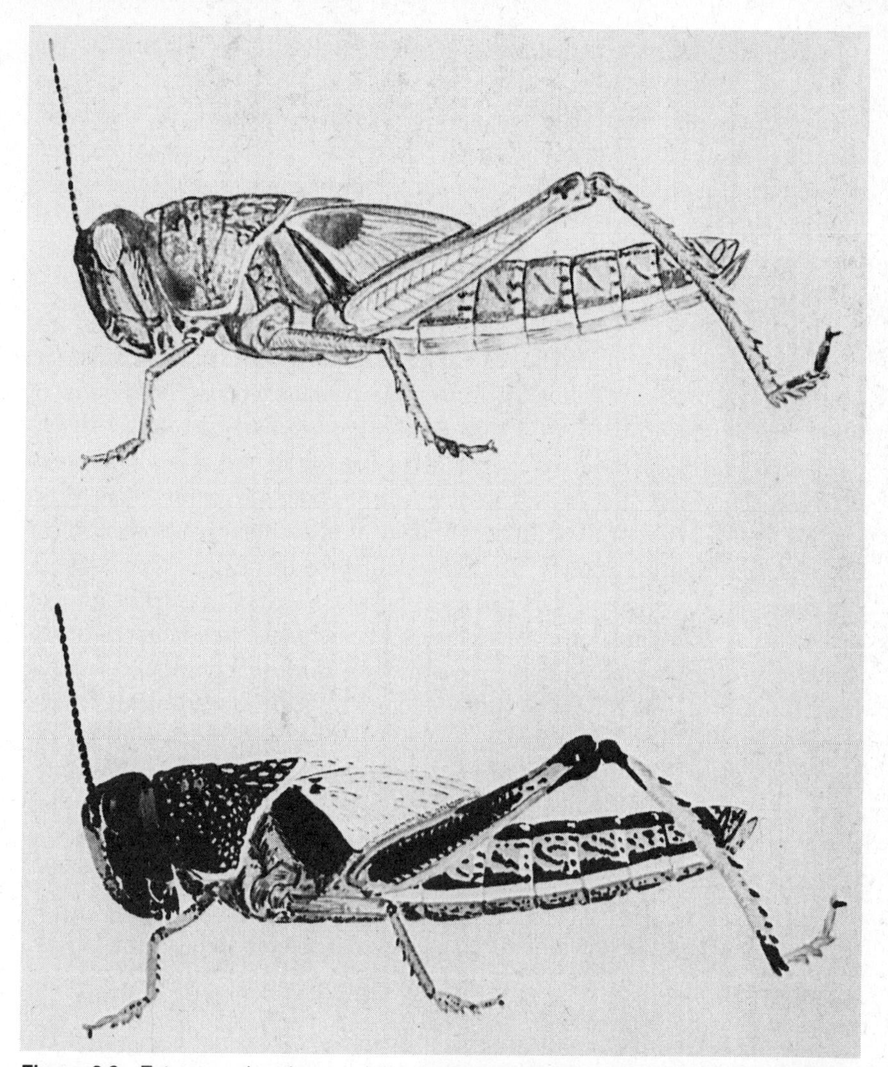

Figure 2.3 Extreme color phases of the desert locust. The pale, greenish form is phase *solitaria* and acts like a solitary grasshopper. The dark form is phase *gregaria* and behaves like a swarming locust. These two forms are so distinct they were believed to be two different species until this century. (Original by Dr. W. Stower, courtesy of the Anti-Locust Research Centre, London)

one generation. Some type of inheritance may thus be involved in phase transformation, but we do not yet know what it is.

The collapse of a locust plague is again largely caused by weather. When moisture falls short, swarms cannot reproduce adequately and green vegetation is in short supply. Thus the most important factor involved in locust population changes, both increases and decreases, is weather (particularly moisture) operating in and through the process of phase transformation.

Not all animal populations go through outbreaks or plagues as the locusts do, and our next example illustrates a single directional change in numbers over 40 years. The red grouse *(Lagopus lagopus scoticus)* is an important game bird in Britain (Figure 2.4). Since 1930 the abundance of red grouse has declined on Scottish moorlands, and research was begun in 1956 to find the reasons for the population decline and to recommend ways of increasing grouse numbers. Red grouse live in open moorlands in Scotland, northern England, and Ireland. Their diet is nearly all one food plant—heather *(Calluna vulgaris)*. They are not migratory but live their entire life within a small part of the moorland. The red grouse year can be divided into two parts: a period of reproductive gain from April to August and a period of overwinter loss from autumn to the start of the next spring breeding season.

Overwinter losses in red grouse occur usually in two stages. Numbers decrease sharply in the autumn, remain stable over the winter, and drop sharply again in the spring. Territorial behavior accounts for these two sharp changes. In the autumn family groups break up because of aggression between the young birds, and two social classes of birds are formed. Territory owners defend a particular area of moor against all comers, and young cocks try to displace territory holders from the previous year. Birds that cannot get territories cannot breed the following spring and are forced to spend most of their time in marginal habitats where heather is scarce. These birds, which form flocks, are then "surplus" birds, excluded by the social system from breeding. Most of the red grouse killed by predators are from these flocks of surplus birds. If a territory owner is killed during the fall or winter, he is quickly replaced by one of the surplus birds.

The second sharp change in the red grouse population occurs in the spring and is controlled by territorial behavior as the breeding season begins.

Figure 2.4 Cock red grouse *(Lagopus lagopus scoticus)* crowing on a lookout rock in his territory. Males occupy territories through most of the year. (Photo courtesy of N. Picozzi)

When grouse are very aggressive and take large territories, numbers fall or remain low. When grouse are less aggressive and take smaller territories, their numbers rise or remain high. What determines how aggressive a male red grouse will be, how big a territory he will take? Two factors seem to be involved. The food available to the mother has a direct effect on the aggressiveness of her offspring. Hens on a high quality diet produce young that are less aggressive and take smaller territories. There is also a genetic component to aggressiveness, so that aggressive cocks produce aggressive chicks.

Breeding success is also strongly affected by nutrition. Moors that are artificially fertilized with nitrogen and phosphorus support heather that is richer in protein and mineral content, and female grouse feeding on this high quality food produce twice as many young as females on poor sites. Good nutrition thus acts in two ways to increase the red grouse population—it increases the production of young birds, and it reduces aggression so that territory size becomes smaller.

The food of red grouse is almost entirely heather, and birds are highly selective in their feeding. They prefer the new shoots of 3-year-old heather plants when they have a choice. The nutrient content of the heather is more important than the amount of heather available. A moor covered with a vast quantity of green heather may be inadequate for red grouse if the plants are too old or the soil too poor. One of the recommendations to Scottish landowners that has come from the red grouse research is to improve heather by controlled burning. Every 12 to 15 years a plot of heather should be burned because by this time the heather has grown too tall for grouse to feed properly. Burnt heather regenerates quickly, and the short nutritious shoots of young heather improve the level of nutrition. In a few years after burning the population size of red grouse goes up. Best results are achieved by rotational burning of long strips 30 meters wide, with older heather left on each side to provide nesting cover. Large fires should be avoided because each grouse's territory must contain nesting as well as feeding areas.

Why did the red grouse population decline during the 1930s? Shooting of red grouse in the autumn has little impact on grouse numbers, and the commonsense idea that too much hunting was responsible for the grouse decline is not correct. There is a large surplus population available every autumn. Many birds shot in the fall are animals that would die anyway during the winter, and the breeding population of territorial birds is not reduced by hunting in the autumn. If a territory holder is shot by a hunter, one of the grouse from the surplus flock takes his territory.

The large decline in grouse numbers since 1930 (Figure 2.5) has been due largely to bad habitat management and overgrazing by sheep. Sheep browse on heather and thus compete with red grouse for food. On moors where proper burning and heather management are practiced, grouse numbers have re-

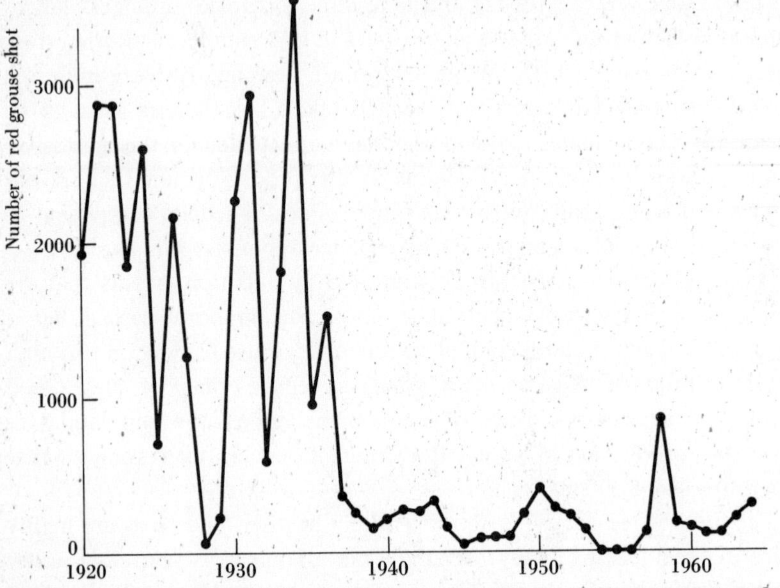

Figure 2.5 Numbers of red grouse shot on grouse moors at Kerloch in eastern Scotland from 1920 to 1964. Short-term fluctuations are superimposed on a long-term decline. (Data from Jenkins and Watson 1970)

bounded to high levels once again. An interplay between nutrition and territorial behavior sets the limits to red grouse populations.

A few mammal and bird populations are unusual in showing regular fluctuations in population size. The famous *lemming* is one that almost everyone has heard about (Figure 2.6). Lemmings are small rodents with very short ears and tail that live in tundra regions across North America and Eurasia. They are famous because they fluctuate dramatically in numbers from year to year, reaching a high density every three to four years only to decline again in a never-ending cycle (Figure 2.7). During the lemming high individuals are everywhere, and the rapid change from scarcity to superabundance has given

Figure 2.6 The Norwegian lemming *(Lemmus lemmus)*. Lemming populations fluctuate violently in three- to four-year cycles in tundra areas of Fennoscandia (where this species lives) and across Siberia and North America. (Photo courtesy of Gunnar Lid, Oslo)

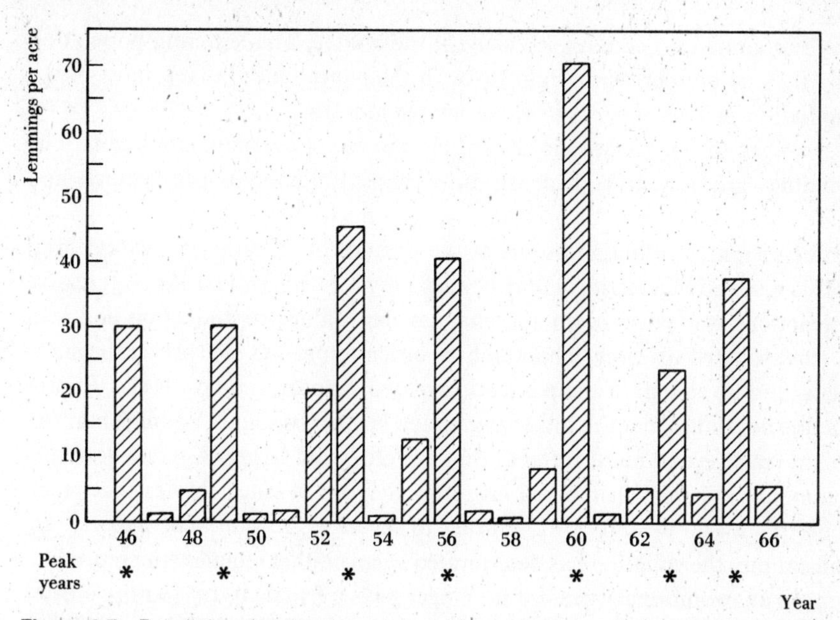

Figure 2.7 Population densities in summer of brown lemming *(Lemmus trimucronatus)* at Point Barrow, Alaska. (After Schultz 1969)

rise to many legends. Lemmings are said to appear by falling from the sky during heavy rainstorms, and to disappear by migrating into the sea in vast herds. Nature is rather less picturesque than the legends would have it, but the statistics are no less dramatic. What does cause lemming populations to rise, reach a high, and then fall? The answer is not completely known, but we think this is what happens.

Begin with a low lemming population in which 1 or 2 lemmings have 4 hectares (10 acres) all to themselves. Conditions are ideal, and reproduction occurs at maximal rates. Females produce a litter of 5 to 8 young every 21 to 25 days. Young lemmings begin to breed when they are 3 to 5 weeks old. But, as lemming numbers increase, more animals come into contact and the amount of strife in the population also rises. An adult male will fight other adult males in the immediate vicinity. The winner obtains access to the breeding females in his home area. The loser often leaves the vicinity to search for a less crowded area. Females defend their nest site and young in the nest against intruders of both sexes.

People have always subjectively viewed the winter as a period of great hardship for animals. But for lemmings the winter may well be the easiest time of the year. Sparse lemming populations will often continue breeding right through an entire winter. As the snow comes, lemmings make tunnels at the snow-ground surface and feed on the frozen buds of tundra mosses and plants. Snow restricts somewhat the free movements of lemmings, and individuals may not come into contact so easily in the winter. The lemming population may thus go on increasing right through the winter and, in fact, most of the population increase occurs in these winter months.

When the snow melts in the spring, the axe of overpopulation falls. The lemmings are now at high densities, perhaps 100 animals per hectare, and aggressive competition for breeding space is maximal. Wounds from fighting can be seen more often and some of the wounds prove fatal. Individuals that are disposessed of a territory may be easy prey for owls and hawks. Aggressive interactions can cause hormonal changes that reduce reproduction and lead to stress-related diseases. The numbers of lemmings fall sharply as the snow melts, and up to half of the population may disappear in a few weeks. Reproduction continues, but the toll of aggression begins to mount. Young lemmings do not reach sexual maturity until they are older and larger. Aggressive adults commonly kill small lemmings just after they are weaned. Males that reach puberty with no wounds are subjected to severe aggression after puberty. By midsummer the situation has deteriorated so much that females stop reproduction. In an evolutionary sense it no longer pays a female to try to raise a litter of young since most of them are killed, and aggression over breeding space has driven the lemming population to the brink.

Once reproduction has ceased, lemming numbers fall rapidly. An accumulation of predators follows the lemming cycle with great interest. Snowy owls are able to reproduce only when lemming numbers are high. Weasels start to breed year round in response to the abundant lemmings they catch. Foxes, hawks, wolves, and other predators all take advantage of high lemming numbers. And these predators all reach their largest numbers just when the lemming population is collapsing under its own internal problems of aggression. So lemming numbers fall through the autumn and the next winter to perhaps a level one-tenth that of the previous summer.

As the snow melts the following spring, lemmings resume reproduction in earnest, but in spite of the relatively low density, only a few lemmings per hectare, they may be unable to reverse the decline. Young lemmings again fail to mature, and the young produced by older females do not survive well. Aggression remains high and signs of wounding are common. The lemmings continue to behave as though they were at high density in a crowded environment when, in fact, they are nearer to low density. As the decline continues through the summer, reproduction again stops, and numbers continue to fall. By this time lemmings are often so scarce that they are difficult to study. They may remain at low numbers for one or two years, and then the cycle begins again. Little is known of what happens during the period of low numbers to make the next cycle possible.

Cycles in populations are not confined to lemmings. Many species of small rodents in the temperate zone of Eurasia and North America also show a three- to four-year fluctuation in numbers. Many of the details are similar to those of the lemming cycle, and all these cycles may have a common explanation.

The lemming cycle is produced by changes in the aggression of individuals in the population, but there is as yet no evidence on whether these shifts in aggression are inherited or not. A limit to population increase is set by aggression. Aggressive individuals do well in crowded populations and continued selection for aggressiveness as crowding increases drives the population to a point where individuals can no longer replace themselves by successful reproduction. Lemmings seem to be an example of a population in which population growth is stopped by the direct and indirect effects of aggression associated with crowding. The control of population does not depend on epidemic disease, bad weather, food shortage, or increased predation.

The natural controls that operate to prevent populations from increasing are never simple, as we have learned to our dismay by interfering with populations. A good way to see this is to look at pest populations. The history of pest control shows how complex natural controls can defeat our best intentions. The conventional idea of pest control is that killing one animal reduces the size of the pest population by one. The arithmetic is simple, the biology is not. Let us see why this commonsense idea is completely wrong.

The wood pigeon *(Columba palumbus)* has been an agricultural pest in Britain for the past two centuries, and the Agricultural Research Council decided during World War II that the pigeon problem could be solved by cutting down the wood pigeon population. The simplest way to do this was to encourage hunters to shoot pigeons. After the war the British government introduced a subsidy to pay half the cost of the cartridges. More than 2 million pigeons were killed each year between 1953 and 1960, and our common sense would predict that the wood pigeon population would decline under this onslaught. But it did not decline, and intensive shooting of wood pigeons both at roosts and at decoys has had no effect on pigeon numbers.

The reason for this violation of common sense is that wood pigeons produce a great excess of young each year. The excess of population is reduced over the winter by food shortage, and the overwinter mortality falls heavily on the younger pigeons. Wood pigeons feed on clover and grain fields in flocks during the winter. Socially subordinate birds, often the younger ones, feed at the front of the flocks, and dominant birds feed at the rear. Subordinate birds try to avoid contact with their neighbors and spend more time nervously looking around to see who is nearby. The result is that subordinate pigeons feed less and thus die first when food becomes scarce.

If winter food supply acts as a natural control on wood pigeon numbers, then shooting birds over the winter only kills birds destined to die from starvation anyway. In fact, shooting could possibly *increase* wood pigeon numbers if shooting reduces the impact of the birds on their winter foods. Paradoxically, *increasing* the numbers of pigeons in the fall might *reduce* the population expected the next spring, if the excess birds destroy the winter food supply more rapidly than would normally occur. If we want to change the numbers of wood pigeons, we must understand and manipulate the natural forces that control their numbers. A great deal of money has been wasted in shooting wood pigeons destined to die anyway.

Of course, the example of the wood pigeon does not mean that overhunting will have no effect on the numbers of other species. Not all populations are as resilient as the wood pigeon. The passenger pigeon was perhaps the most abundant bird in eastern North America when European colonists first arrived. It is now extinct, destroyed by a combination of overhunting on an excessive scale during the 1800s and habitat destruction associated with farm clearance. The last passenger pigeon died in the Cincinnati Zoo in 1914.

Attempts to limit the European rabbit by eliminating individuals is another case in which conventional ideas of pest control failed. The European rabbit has been a pest not only in Australia (see Chapter 1), but also in Britain and Europe. Commercial rabbit trapping actually caused rabbit populations to increase rather than decline. By removing 30 to 40 percent of the breeding stock commercial trappers succeeded in undermining the social structure of the rabbit population. Trapping selectively removes the dominant animals because they move about more. This reduces territorial aggressiveness in the rabbit population and enables subordinate females to breed. Since there are more subordinate than dominant females, the total production of the rabbit population goes up even though trapping is removing animals. As with the wood pigeons, the pest problem may be *worse* after artificial controls are started.

Many people persist in the belief that we can control and indeed *eliminate* pests by killing them in large numbers. Control may be possible, but elimination rarely is. The confusion of killing and controlling has led us into the use of an array of toxic chemicals for pest management. Fortunately, the storm over ecological disruptions caused by pesticides, first exposed to the public in Rachel Carson's *Silent Spring,* has stimulated a more sophisticated approach to pest control. To control a pest we must understand the ecological processes that act as natural controls, and we must avoid interfering with these. Pests are pests because they are cleverly adapted to the human environment or because they are released from natural controls by human intervention. Pest populations resemble rubber balls—we can knock them down by brute force, but they rebound with endless resilience. To realize this is one mark of ecological wisdom.

Not all species are pests of course and ecologists recognize that there are many endangered species that we can too easily control and eliminate—the

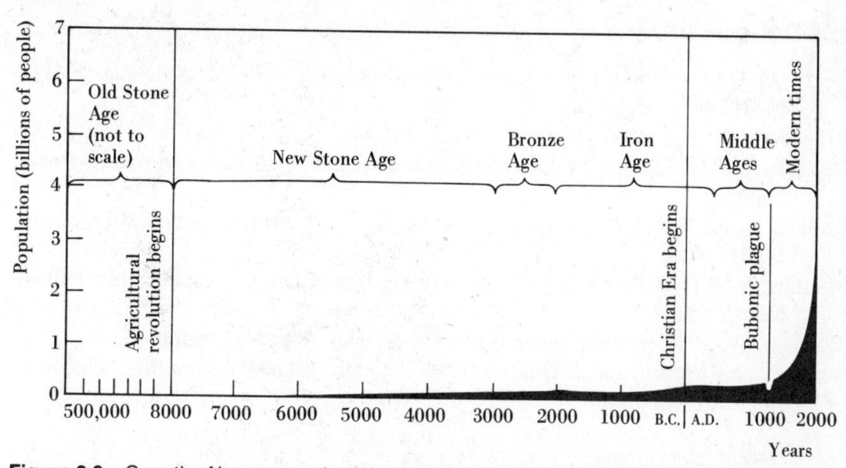

Figure 2.8 Growth of human numbers for the past 500,000 years. If the Old Stone Age were in scale, its base line would extend about 5 meters to the left. (After *Population Bulletin* 18, No. 1)

giant panda, the whooping crane, the California condor, to mention only a few. The larger the animal, the more likely it is that humans can control or eliminate it if they so choose.

If populations do not increase without limit, the human population can be no exception. Throughout most of the last 500,000 years humans have been a "rare" species on the earth (Figure 2.8). Human populations rose and fell back again under the combined assault of starvation, disease, and climatic catastrophes added to the self-inflicted losses caused by local and regional wars. Population growth, when it occurred, averaged less than 0.01 percent per year. Egypt, for example, had virtually the same number of people in 2500 B.C. as it had in A.D. 1000. Gradual improvements in agriculture, public health, and housing and a reduction in warfare over the last thousand years have put the human population on a growth spiral that has become extremely rapid during this century.

The human population cannot increase without limit. Most people accept this fundamental truth, and the problem comes down to how and when the human population will stop growing. We differ from other organisms in having the capability to impose our own controls, rather than relying on natural mechanisms of famine, plagues, or aggression to set a balance. A major benefit of human civilization has been a reduction in deaths caused by disease and starvation, enabling individuals to live a longer life. But we must reduce the birth rate to a low value equal to the death rate in order to stop our population increase. Like all animal populations, we can persist either with a high birth rate equal to a high death rate or (as most would prefer) with a low birth rate equal to a low death rate. This choice is upon us now. This is the single most important problem we face in the world today.

FURTHER READING

Coale, A. J. 1974. The history of the human population. *Scientific American* 231(3): 40–51.

*Crowcroft, P. 1966. *Mice All Over.* Foulis, London.

DeBach, P. 1974. *Biological Control by Natural Enemies.* Cambridge University Press, London.

Ehrlich, A. H. 1985. The human population: size and dynamics. *American Zoologist* 25: 395–406.

Malthus, T. R. 1798. *An Essay on the Principle of Population.* Reprinted by Macmillan, New York.

Marsden, W. 1964. *The Lemming Year.* Chatto and Windus, London.

Moss, R., A. Watson, and J. Ollason. 1982. *Animal Population Dynamics.* Chapman and Hall, London.

*Highly recommended.

chapter 3

Good and Poor Places Exist for Every Species

Nature is a mosaic of good, poor, and impossible habitats for any beast or plant, and we now ask how one can tell the good from the poor. Every bird watcher, mushroom picker, hunter, or fisherman knows some places where you can be sure of finding individuals of some desirable animal or plant and other places where you would have to search much longer to locate an individual. Ecologists try to find out why one habitat is better than another for two practical reasons. Conservation programs are often devoted to saving endangered species and protecting established species. We can protect a desirable species, like the bald eagle or caribou, only if we know what a "good" habitat for that species includes. Pest control programs are the reverse side of this coin. If we know the species' ecological requirements, we can make habitats "poor" for pest species.

A species is affected by the environmental factors of weather, nutrients, other species, and shelter. Continued presence in an area depends upon *all* these factors being favorable. All the links in the chain must be present for the species to thrive, and the ecologist tries first to locate the weak links, realizing that the complete description of the whole chain may be impossible for the present time. Any environmental factor may provide the weak link. Consider a few examples.

The pied flycatcher *(Ficedula hypoleuca)* is a black and white bird common in woodland areas of Europe. It nests in tree holes and feeds largely on insects. Because most of the old trees in Europe have been cut for lumber, holes are in short supply and the distribution and abundance of the pied flycatcher can be influenced by putting up nest boxes. For example, in southern Finland pied flycatchers are uncommon in spruce forests and do not occur at all in pine forests, but the number of birds increases if nest boxes are placed in the woods.

	No nest boxes (1961)	**Nest boxes added (1962)**
Spruce forest	3 pairs	52 pairs
Pine forest	none	85 pairs

The density of pied flycatchers can be manipulated by the spacing of the nest boxes (Figure 3.1). This situation seems to be remarkably simple: The more nest boxes, the higher the density of breeding flycatchers in pine, spruce,

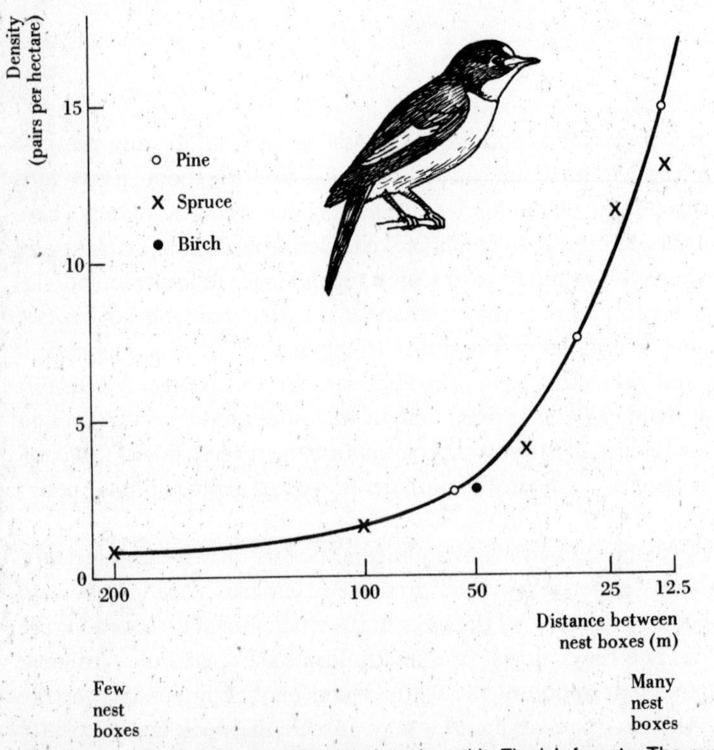

Figure 3.1 Pied flycatchers *(Ficedula hypoleuca)* in Finnish forests. The number and density of birds can be manipulated by supplying artificial nest boxes. The more nest boxes, the more birds. The closer the boxes, the greater the density of population in an area. This happens in pine, spruce, and birch forests. (Data from von Haartman 1971)

and birch forests. There is undoubtedly some upper limit to how many nest boxes and breeding pairs you could have in a forest, but this upper limit is high indeed. Nesting places are a limiting resource for the pied flycatcher, and by putting up nest boxes, that is, by supplying the limiting resource, you can change a poor place into a good place for this bird species.

Kangaroos and Australia are synonymous for most people, and the abundance of the large kangaroos has gone up since the British colonized Australia. The increase in kangaroo populations has occurred in spite of intensive shooting programs, since kangaroos are considered pests by ranchers and are harvested for meat and hides. The reason seems to be that ranchers have improved the habitat for the large kangaroos in three ways. First, in making water available for their sheep and cattle, the ranchers have also made it available for the kangaroos, removing the impact of water shortage for kangaroos in arid environments. Second, ranchers have cleared timber and produced grasslands for livestock. Kangaroos feed on grass, and so their food supply has been increased as well as the water supply. Third, ranchers have removed a major predator, the dingo *(Canis familiaris dingo)*. The dingo is a doglike predator, the largest carnivore in Australia. Because dingoes eat sheep, ranchers have built some 9660 kilometers of fence in southern and eastern Australia to prevent dingoes from moving into sheep country. Intensive poisoning and shooting of dingoes in sheep country, coupled with the dingo fence that prevents recolonization, has produced a classic experiment in predator control. Figure 3.2 gives the results for red kangaroos *(Macropus rufus)*. There is a spectacular increase in the abundance of red kangaroos when dingoes are eliminated. Densities of kangaroos are 166 times higher in New South Wales

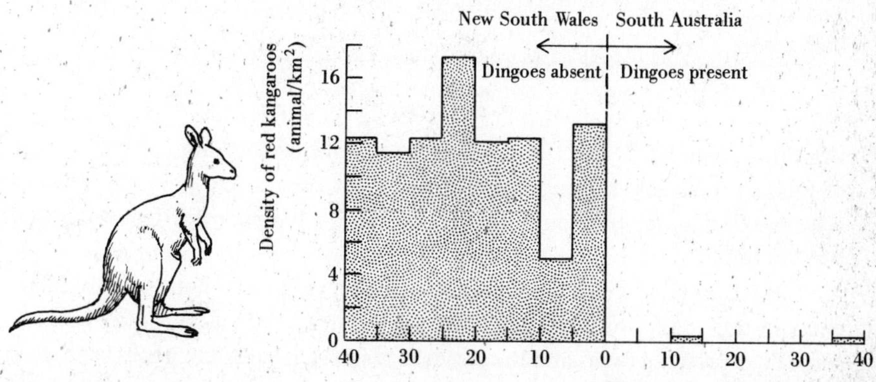

Figure 3.2 Density of red kangaroos *(Macropus rufus)* on a transect across the New South Wales–South Australia border in 1976. The border is coincident with a dingo fence that prevents dingoes from moving from South Australia into the sheep country of New South Wales. (After Caughley et al. 1980)

than in South Australia. Dingoes are able to hold kangaroo numbers low in South Australia because the dingoes are not solely dependent on the kangaroos as food supply. They have alternate prey such as rabbits and rodents to sustain them. Dingoes eat sheep and, by getting rid of one problem, the ranchers have helped to accentuate the kangaroo problem.

The improvement of the food supply is a powerful technique for turning a poor habitat into a good one. Much of wildlife and fisheries management operates on this assumption. The red grouse *(Lagopus lagopus scoticus)* is a good example (see Chapter 2). Grouse habitats range from good moors which support high densities of grouse and big shooting bags to poor moors which support much lower grouse densities and little shooting. Red grouse eat heather and the quantity, age, and nutrient quality of the heather are related to grouse density. Since grouse prefer young heather, good moors have more patches of heather of younger age than poor moors. However, since grouse need old heather for cover, the best moors are a patchwork of old and young heather. Some of the worst moors have nothing but young heather. Moors on good soil produce heather that contains more nitrogen and phosphorus than heather from poor soil. Thus improvements in heather increase the red grouse stocks on a moor, and the agricultural generalization—good food, more animals—holds as well for these wild birds.

Mountain hares *(Lepus timidus)* also feed on heather and they also have high numbers on good heather moors.

	Mountain hares (no./km²)	**Red grouse (no./km²)**
Poor moors	4	60
Rich moors	51	119

Both mountain hares and red grouse are selective feeders and choose heather of high protein content on both rich and poor moors. On poor moors even selective feeding is not enough to compensate for the poor nutrient quality of the food plants.

Deer populations in North America respond in a dramatic way to differences in the quality of nutrition. On a good range production is high because some does produce twins and the survival of fawns is high. On poor range does twin less frequently, and fawns grow slowly, entering their first winter at a smaller size. Slow growth rates also increase the age at sexual maturity. On poor range does do not produce their first fawn until 3 years of age, but on good range most produce young at 2 years of age. Disease and parasite loads may also be higher in undernourished deer. Male deer have a higher mortality rate than female deer, and this difference is accentuated on poor range so that females come to outnumber males greatly on poor range.

Deer management thus aims at providing places with an abundance of food plants of high nutrient content, but this is easier said than done. By feeding selectively, deer reduce the abundance of the high quality food plants, and if deer are allowed to reach high numbers, they can rapidly turn a good range into a poor one.

Forest fires sometimes cause habitat alterations on a grand scale, and a whole group of animals and plants exploits these situations. In the Pacific Northwest blue grouse *(Dendragapus obscurus)* populations increase dramatically after a fire destroys the mature conifer forest (Figure 3.3). Similar increases occur after logging has occurred, so the fire itself is not as important as the opening up of the habitat. Open habitats provide more nesting sites for blue grouse and more food for their chicks. As the forest slowly regenerates, blue grouse stay at high densities and hunting success is good. But once the conifers become dense, blue grouse begin to return to low numbers, and hunting success becomes poor.

Moose are like blue grouse in their response to fire and logging. Moose feed on many of the broad-leaved plants, shrubs, and small trees that colonize burned or logged areas, and as these plants are eliminated in the dense forest,

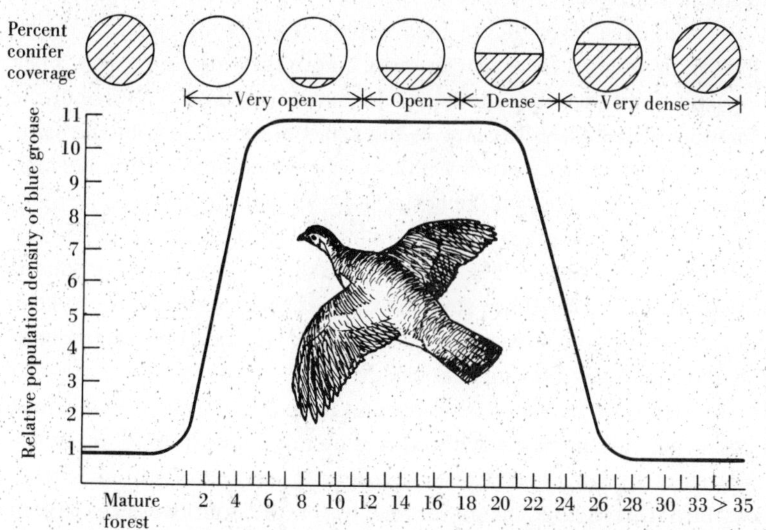

Figure 3.3 A generalized curve for growth, stability, and decline of blue grouse *(Dendragapus obscurus)* populations following logging or burning of the mature forest in western British Columbia. Population density is low in the mature forest, but rapidly increases to high density after the forest is cleared. (After Redfield et al. 1970)

the moose decline in numbers. Species of this sort are sometimes called *fugitive species* because they move as environments change and depend on the opening up of new habitats by fire, logging, or other disturbances to remain at high numbers. Fugitive species can be very successful. Moose are extremely abundant in Sweden because logging continually provides open habitats with the shrubs moose eat.

Some plants depend upon fire to maintain high densities. The most famous example is the giant sequoia *(Sequoiadendron giganteum)* of California. These magnificent trees are replaced by other conifers, but only in the absence of fire. Conservation attempts to protect the sequoia forests by stopping all forest fires have, in effect, almost doomed these trees to disappear, and attempts to restore fire to a useful place in forest management are currently under way in the National Parks Service of the United States (Figure 3.4). Large sequoia trees have a thick, fire-resistant bark and so they are not damaged by ground fires that are fatal to many other conifers, such as white fir and sugar pine. Sequoia seedlings also germinate best on bare mineral soil, and ground fires provide a good environment for seedling establishment by removing the litter on the forest floor. Thus organisms as different as blue grouse, moose, and sequoia trees may all depend upon habitat changes brought on by fire in order to keep their numbers high. Good habitats are not necessarily those that are never disturbed.

Manipulation of the structure of the habitat can affect the numbers of animals using an area, even if the food supply itself is unchanged. Young salmon fry live in streams for a year or more before they migrate to sea. They take up feeding territories that they defend against intruders. The size of the feeding territory depends on the size of the fish, but among fish of the same age one can change territory size by altering the topography of the bottom. When large stones are placed on the bottom of a stream, the fry become more visually isolated from one another; as a result territories become smaller, and so population density goes up (Figure 3.5). The same effect can be produced by increasing the current. Higher currents cause the small salmon fry to keep closer to the bottom of the stream, increasing visual isolation between individual fish in the same way that large rocks do.

Habitat structure may also be important for wildlife species, and it may be possible to increase numbers without adding any more food to an area if we provide more shelter. Scaled quail *(Callipepla squamata)* are an important game bird throughout semiarid regions of the southwestern United States. In southeastern Colorado the numbers of scaled quail can be increased by increasing the amount of cover for resting. Resting cover can be provided cheaply by constructing brush piles of dead tree limbs, old posts, and old Christmas trees, but the quail use old car bodies and abandoned

Figure 3.4 Prescribed burning in 1969 in the Redwood Mountain Grove of giant sequoias, Kings Canyon National Park, California. Fire consumes the accumulation of forest fuels, leads to a recycling of nutrients and reduction in wildfire hazard, and prepares a seedbed for sequoias. During these early efforts National Park Service crews used fire hoses as an added safety precaution. (National Park Service photo courtesy of Bruce M. Kilgore)

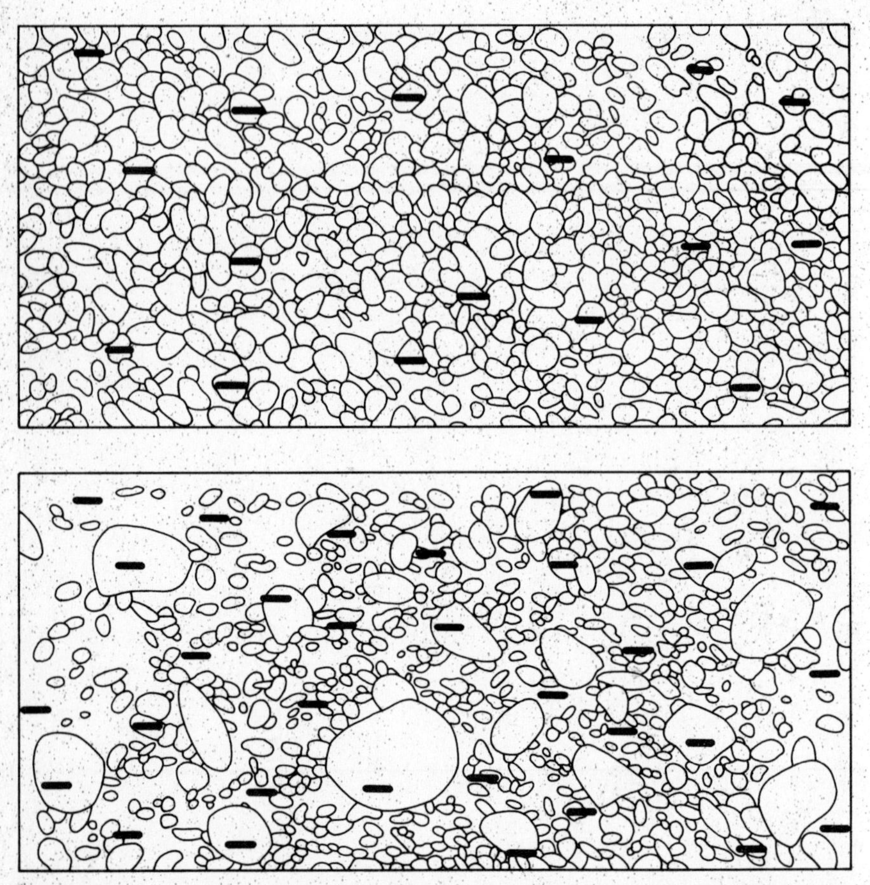

Figure 3.5 The position of Atlantic salmon fry within territories in two sections of an artificial stream with and without large boulders. Salmon territories are reduced by half when boulders are present, and thus more fry are able to live in the lower section. (After Kalleberg 1958)

farm machinery as well. Quail use resting cover for protection from wind, weather, and bird predators, and the provision of resting cover seems to change poor quail habitat into better habitat and to increase quail numbers on abandoned farmlands.

	Number of scaled quail in winter	
	Area A (no brush shelters)	**Area B (brush shelters added)**
Before experiment		
1961–1962	13	0
After experiment		
1962–1963	14	29
1963–1964	19	108
1964–1965	0	6

A severe drought in 1964–1965 reduced quail numbers on all areas because of food shortage, but areas with shelter added maintained higher numbers of quail.

Pests are species that exist at densities that interfere with our own activities. The best methods of pest control are based on techniques for changing good habitats into poor ones for the pest species. The attempt to create bad habitats by introducing chemicals has not been successful in the long run. Because many insect predators feed on insect pests, one of the first surprising results of chemical spraying against insect pests is that spraying can actually *increase* the abundance of a pest! Figure 3.6 shows one example in which lemon trees became severely infested with scale insects after a DDT spraying. Because chemical sprays kill all insects, they destroy both the bad ones (pests) and the good ones (insect predators that eat the pests). Unless every single pest individual is killed, the pest can increase after the spraying in the absence of any predators.

Pest control has also become a difficult art because many of our most serious pests have become genetically resistant to chemical sprays that were formerly lethal. A good habitat cannot be converted into a poor one simply by adding a chemical spray. We must look at the problem more ecologically.

In some cases the addition of an insect predator has reduced a pest

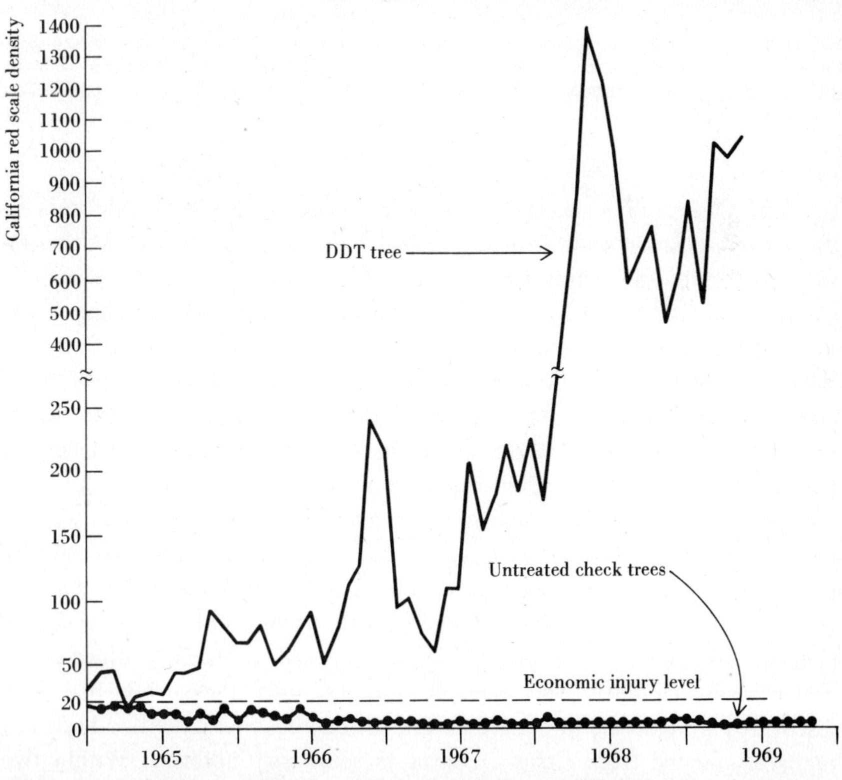

Figure 3.6 Increases in California red scale *(Aonidiella aurantii)* infestation on lemon trees caused by monthly applications of DDT spray. Nearby untreated lemon trees suffered no damage because a variety of insect parasites and predators kept the red scale under biological control. (After DeBach 1974)

Figure 3.7 An illustration of crop resistance to insect attack. Rice that is resistant to attack by the rice stem borer is shown on the left of this photo, and rice that is susceptible is shown on the right. (Photo courtesy of Dr. M. Pathak, International Rice Institute, Los Banos, Philippines)

problem. One striking success in this field has been achieved in California on the cottony-cushion scale *(Icerya purchasi),* a scale insect that sucks the sap from the leaves and twigs of citrus trees. This insect was first discovered in California in 1872, and by 1887 the whole citrus industry of southern California was threatened with destruction. In 1888 the U.S. government sent Albert Koebele of the Division of Entomology of the Department of Agriculture to Australia, the native home of the cottony-cushion scale to search for insect predators that might attack the scale in California. Koebele sent two insect species back to California, one a small parasitic fly and the other a predaceous ladybird beetle called the vedalia *(Rodolia cardinalis).* By January 1889 a total of 129 ladybird beetles had been released near Los Angeles under an infested orange tree covered with a large tent. By April 1889 all the cottony-cushion scales on the tree had been destroyed, and the tent was then opened. By June 1889 vedalia beetles had been distributed to other citrus orchards. By October 1889, scarcely one year after the vedalia had been discovered by Koebele in Australia, the cottony-cushion scale had been virtually eliminated from citrus orchards in southern California. Within two years it became difficult to find a single specimen of the scale on citrus trees.

This operation cost only $1500, and saved millions of dollars every year. Citrus orchards that contained vedalia were no longer good habitats for the cottony-cushion scale.[1]

Pests cannot always be reduced by the introduction of new predators or parasites, and some of our worst pest problems, like the corn borer in the central United States, persist in spite of the introduction of many predator species. For these pests we have to use other methods, and one of these is to try to develop crop plants that are resistant to pests (Figure 3.7).

One of the most spectacular examples of crop resistance put to practical use was the resistance of the American grape plant *(Vitis labrusca)* to the root aphid grape phylloxera. The European grape plant *(Vitis vinifera)* is highly susceptible to phylloxera, which kills the roots. The phylloxera was accidentally introduced to France in 1861, and by 1880 the French wine industry was near collapse from the destruction caused by it. By a massive program of grafting susceptible vines of the desirable European grape plants to resistant rootstocks of the American grape, the French wine industry was saved. The grape phylloxera is now scarce and rarely a problem, and the French wine industry flourishes.

The Hessian fly was a serious pest of wheat in the United States until resistant varieties of wheat were developed. The larvae of the Hessian fly secrete toxins and form galls at the base of the leaf of the wheat plant; they then feed on plant fluids in the gall. Gall formation slows the growth of the wheat leaves and may kill the whole stalk of wheat. If the wheat plant survives, the yield of grain is reduced 25 percent or more, and some stalks break or bend at the point infested by the Hessian fly larva. A breeding program for resistant varieties of winter wheat was begun as early as 1880. Several genetic characteristics were identified in the 1920s and 1930s for Hessian fly resistance. Some varieties of wheat were found to cause heavy mortality of Hessian fly larvae through some chemical antibiotic interaction. Others were found to have stronger stems that do not weaken under fly infestation. As a result of this work, a resistant variety of wheat called Pawnee was developed, and as more and more Pawnee wheat was planted in Kansas the population of the Hessian fly declined to virtual extinction. The wheat field had been changed from a good habitat for Hessian flies to a poor one, and the Hessian fly was a pest no longer.

There is a danger in the development of genetic strains of crops that are highly resistant to particular crop pests—they can be a victim of their own success! A genetically uniform crop strain that is resistant to a particular

[1]Ironically, the cottony-cushion scale reappeared with the advent of DDT. DDT eliminated the vedalia beetle from some local areas and the scale populations increased to damage levels.

organism may be propagated widely and come to occupy large areas. If a pest appears to which the particular crop strain is not resistant, an epidemic can take place. In 1970 exactly this occurred in the corn belt of the United States when a fungus disease called Southern corn leaf blight suddenly appeared in a new strain well adapted to hybrid corn. The fungus swept through the corn belt and destroyed about 15 percent of the U.S. corn crop. The hybrid corn that most farmers were using in 1970 was not resistant to Southern corn leaf blight. A warning was sounded—a mixture of crop strains may be better than a single strain planted everywhere. Genetic simplification of crop plants is a most unwise policy in a world filled with parasites, diseases, and predators. Not all the eggs in one basket, please!

Humans can artificially select crop plants that are resistant to particular pests, but this type of selection also occurs in nature. The production by a plant of chemicals that discourage feeding is an adaptive mechanism that may be selected for over evolutionary time. Ecologists are just starting to appreciate the vast array of chemical interactions that occurs in natural populations. Plants contain a variety of chemicals which have always puzzled plant physiologists and biochemists. These chemicals are found in some plants and not in others, and many are familiar to us already. The spices cinnamon and cloves are two examples. Nicotine is an alkaloid found in tobacco plants, and the active ingredient of marijuana is a terpene produced by hemp *(Cannabis sativa).* Caffeine is another alkaloid plant chemical. A great variety of these plant chemicals are found in different species.

These chemicals seem to have been evolved by plants primarily as a defense against herbivores. Plants that contain a particular chemical in their leaves could be favored by natural selection if they are eaten less by animals. Of course, animals do not sit idly by while plants evolve defense systems. Herbivores can circumvent the plants' defenses either by evolving enzymes to digest and detoxify plant chemicals or by timing their life cycle to avoid the noxious chemicals of the plants. Consider one example.

The English oak *(Quercus robur)* is a dominant tree in the deciduous forests of western Europe, and its leaves are eaten by the larvae of over 200 species of moths and butterflies. The attack of insects is concentrated in the spring of the year, because oak leaves become less suitable food as they age. Caterpillars of the winter moth, for example, will grow well if they are fed oak leaves from the middle of May, but will grow poorly and die if they are fed oak leaves from early June. Oak leaves in the spring become rapidly darker in color and tougher to chew. At the same time the amount of tannin in the leaves increases, and the amount of protein decreases. Tannins are a group of plant chemicals that discourage animal feeding. They form complexes with proteins in the gut, reducing the digestibility of proteins. Thus the oak has defended itself chemically against herbivores, and the herbivores have responded by concentrating their feeding on the young leaves in the spring of the year. If we could artificially produce an oak tree that had no tannins in its leaves, it could presumably be devoured by herbivores at other seasons.

Another method for turning good habitats for pests into poor ones is diversified planting. A dominant image of modern agriculture is an endless field planted to a single crop, contrasting greatly with the smaller fields and more diversified farming practices of a less mechanized day. Some pest problems may be produced by these extensive monocultures because it provides for pests a concentration of suitable habitats rather than the mixture of suitable and unsuitable habitats that characterizes the natural world. Small fields surrounded by hedgerows or fence lines may support lower pest populations. Insect predators can live in the hedgerows and move into the crop field once the crop begins to grow. Thus a pest population may be held at low numbers by its predators. Large fields may have few sources of insect predators so that pest populations can explode before natural control agents ever reach the scene. Unfortunately, the suggestion that agriculture return to a small scale of fields with different crops interspersed is difficult to implement in western societies that prize large-scale automation and maximum production per work-hour in agriculture. The Chinese have adopted the small-scale type of agriculture with interspersed crops that is rarely seen in western countries. We do not yet have a good appreciation of how habitat interspersion and the spatial scale of a species' movements interact to result in good or poor habitats. Consequently, we are unable to put a precise monetary value on adopting one type of agricultural system over another.

The most significant message that we should retain from this analysis is to stop our attempts to solve pest problems by looking for methods of killing animals and to use instead our ecological knowledge of the pest *to change good habitats for the pest into poor habitats.* For the last 50 years we have largely used the first approach. The challenge of the next 50 years is to solve these problems with greater success, at less cost, and with less environmental damage by developing ecological tricks that exploit the weak links in the pest's life chain. By proper habitat management we can both assist the species we wish to conserve and reduce the species that act as pests.

FURTHER READING

Bailey, J. A. 1984. *Principles of Wildlife Management.* Wiley, New York.

Clawson, M. 1975. *Forests for Whom and for What?* Johns Hopkins Press, Baltimore, Md.

Conway, G. R. (ed.) 1984. *Pest and Pathogen Control: Strategic, Tactical, and Policy Models.* Wiley, New York.

Ehrenfeld, D. W. 1972. *Conserving Life on Earth.* Oxford University Press, London.
*Leopold, A. 1949. *A Sand County Almanac.* Oxford University Press, New York.
Rosenthal, G. A. 1986. The chemical defenses of higher plants. *Scientific American* 254(1): 76–81.
*van den Bosch, R. 1978. *The Pesticide Conspiracy.* Doubleday, New York.
Woods, A. 1974. *Pest Control: A Survey.* McGraw-Hill, New York.

*Highly recommended.

chapter 4

Overexploited Populations Can Collapse

Humans have harvested natural populations of animals and plants from the earliest days. When harvesting was light, no problems arose, and local depletions could always be avoided by moving to a new place. As the human population increased and new technologies were developed, harvesting became more intensive and larger areas were affected. The destruction of forests in the Mediterranean region was nearly completed by Roman times, but in spite of these early problems, only during the last century have scientists attempted to analyze the overexploitation problem in an ecological perspective.

By the late 1800s people began to see examples of overfishing in some marine fisheries. In 1885 the Dalhousie Committee was set up in the United Kingdom to investigate the alleged depletion of fish stocks by the trawl net and the beam trawl. The committee was unable to provide an answer to the problem because no ecological data were available on the fish stocks of the North Sea, and so they recommended that the government begin to collect adequate fish statistics and start scientific work on fish population dynamics.

From 1890 to 1930 fisheries scientists began to sort out the pieces of the puzzle that were needed to analyze the overfishing problem. In 1898 C. Hoffbauer discovered that the age of many fish can be read from growth rings on the scales. This discovery meant that scientists could determine the age composition of a fish population. In 1918 A. G. Huntsman in Canada and F. I. Baranov in Russia independently found that fishing changes the age composition of a fish stock. Fishing removes the older fish, which are also the larger fish, and when a fishery is just starting up, the catches of these larger fish may be very good indeed.

In 1931 the British scientist E. S. Russell published a theoretical analysis of the overfishing problem which set the exploitation problem in a clear light. For a fishery, interest usually centers on the *weight* of the catch rather than on the *number* of fish, and so Russell analyzed the weight dynamics of an exploited population. Two factors decrease the weight of the fish stock during the year—natural mortality and fishing mortality. Fishing mortality is just the fisherman's catch, and in any exploitation problem we view humans as just a large predator removing animals from the population. Natural mortality includes all the losses caused by natural agents—disease, parasitism, and natural predation. Similarly, two factors increase the weight of the stock—recruitment of young fish and growth of older fish. *Recruitment* is the addition of new animals to the catchable population (or *stock*) of fish.

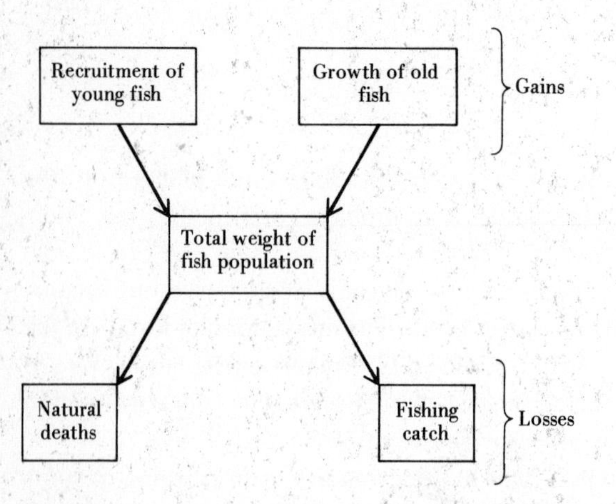

To balance the fish population, Russell pointed out, gains must equal losses, so that

Recruitment + growth = natural mortality + fishing mortality

A critical problem now appears. Before fishing began, recruitment plus growth, on the average, must have been equal to natural mortality if the population was stable. Now, with exploitation, this equation becomes unbalanced and the total weight of the fish population begins to fall until either recruitment or growth goes up or natural mortality goes down. Unless one or more of these three elements changes, the exploited population will become extinct.

An elegant demonstration of what happens in an exploited population

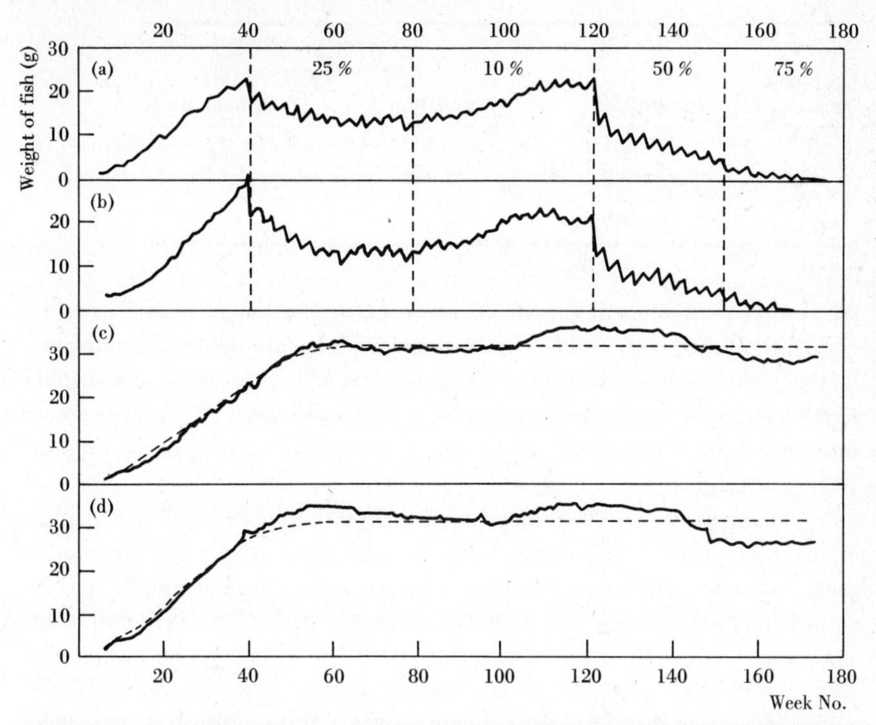

Figure 4.1 Population biomass changes in guppies *(Lebistes reticulatus)* maintained in the laboratory. Two control populations (c and d) are not exploited. Experimental populations (a and b) are subjected to cropping after week 40 at the indicated rates. (After Silliman and Gutsell 1958)

is provided by Silliman and Gutsell (1958) who raised guppies *(Lebistes reticulatus)* in laboratory aquaria. They maintained two populations as unmanipulated controls and two populations as experimental "fisheries" subjected to a sequence of four rates of fishing (Figure 4.1). Populations were counted each week and cropped every third week. For example, a 25 percent cropping rate means that every fourth fish was removed from this population during the count at week 40, 43, 46, and so on.

Control guppy populations reached a stationary plateau around 32 grams of biomass by week 60 and remained at this level for the duration of the experiment. Cropping at 25 percent tri-weekly reduced the experimental populations to about 15 grams. A reduction of the cropping to 10 percent increased both experimental populations to about 23 grams, and the imposition of a 50 percent cropping rate in week 121 caused a decline in population weight to about 7 grams. A cropping rate of 75 percent was too great for the fish to withstand, and both experimental populations were driven extinct by this severe overfishing.

By weighing the guppies removed at each cropping, Silliman and Gutsell could determine the yield to the "fishery."

Cropping rate (%)	Weeks	Average weight of yield per cropping (g)	
		Population A	Population B
10	100–118	2.20	2.51
25	61–76	3.35	3.58
50	136–148	3.88	3.58
75	163–172	0.82	0.40

Maximum yield occurred at the 50 percent cropping rate, but as Figure 4.1 shows, the guppy population continued to decline at this high rate of cropping, so this yield was not sustainable. The best guess is that the maximum sustainable yield could be obtained at exploitation rates between 30 and 40 percent at a population biomass of 8 to 12 grams, compared with 32 grams in unexploited controls.

The message of this simple experiment with guppies is clear and illustrates three principles of exploitation. (1) *Below a certain level of exploitation populations are resilient and increase survival, growth, or recruitment to compensate for fishing loss.* (2) *Exploitation rates may be raised to a point at which they cause extinction of the resource.* (3) *Somewhere between no exploitation and excessive exploitation there is a level of maximum sustainable yield.* Let us examine a few cases of exploitation to see some of the difficulties of applying this simple theory to the real world.

The Peruvian anchovy *(Engraulis ringens)* lives in an area of upwelling of cool, nutrient-rich water along the coasts of Peru and northern Chile. The upwelling causes high productivity in the coastal zone. The Peruvian anchovy is a short-lived fish, spawning first at about 1 year of age and rarely living beyond 3 years. It is small, about 12 centimeters at 1 year and seldom reaching 20 centimeters in length. Young anchovy enter the fishery at only 5 months of age (8 to 10 centimeters). Anchovy occur in schools and are caught near the surface.

The Peruvian anchovy fishery was the largest fishery in the world until 1972, when it collapsed. From 1955, when the major fishery first began, the anchovy catch doubled every year until 1961. In 1970, 12.3 million metric tons (1 metric ton = 1000 kilograms or 2205 pounds) were harvested, and this single fishery comprised 18 percent of the total world harvest of fish. Anchovy are taken both by fishermen and by large colonies of seabirds, and these two sources of mortality must be estimated to get total losses. Figure 4.2 shows the fishery catch of anchovy from 1955 to 1981. The highest anchovy catch, 12 to 13 million metric tons in 1970, was above the estimated maximum sustainable yield of 9 to 10 million tons. From 1964 to 1971 the catch was close to the estimated maximum sustainable yield, and this fishery seemed to be a good example of a scientifically managed population. But hidden in all these figures

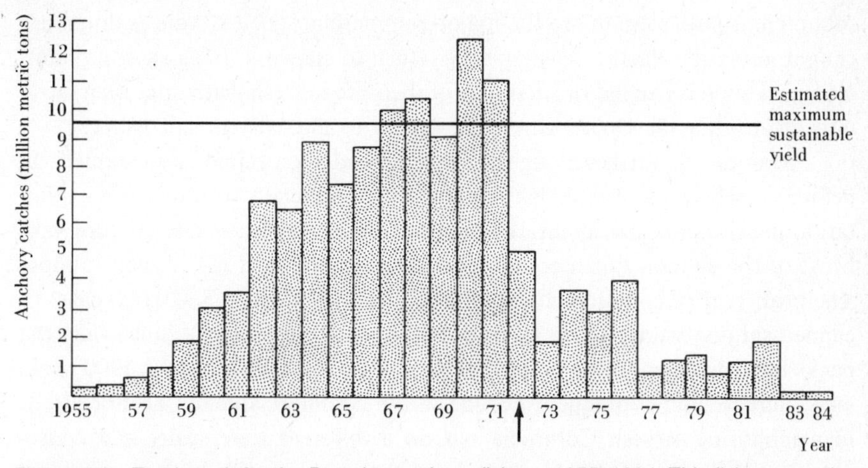

Figure 4.2 Total catch for the Peruvian anchovy fishery, 1955–1984. This fishery was the largest in the world until it collapsed in 1972 in association with El Niño (arrow). In spite of reduced fishing there has been no recovery since then. (Data from M. H. Glantz)

is a critical assumption—that the environment does not change. Behind all the fisheries management decisions was the simple assumption that average environmental conditions would prevail so that the anchovy population would grow, reproduce, and survive as well in the 1970s as it did in the 1960s.

In 1972 average conditions disappeared, and the Peruvian anchovy fishery collapsed. Early in 1972 the upwelling of cold, nutrient-rich water off the coast of Peru weakened, and warm tropical surface water moved into the area. This phenomenon—known as El Niño, because it often happens around Christmas—occurs about every seven years and greatly changes the ecology of the area. The productivity of the sea drops because the warm surface water is poor in nutrients, seals and seabirds starve, and anchovy move south to find cooler waters. In early 1972 very few young fish were found; the spawning of 1971 had been poor, only one-seventh of normal. Adult anchovy were highly concentrated in cooler waters in early 1972, and these concentrations produced large catches for the fishermen. By June 1972 the anchovy stocks had fallen to a low level, catches had declined drastically, and no young fish were entering the population. The fishery was suspended to allow the stocks to recover, but since 1972 there has been no return of the anchovy toward its former abundance and populations of seabirds have also remained low. The economic consequences of the fishery collapse have been great. Some of it might have been avoided if the fishery had been closed a few months earlier, or if the fishing intensity had been less than the maximum 9 to 10 million metric tons per year.

The decline of the Alaska salmon fishery is a good example of overfishing applied to an apparently "inexhaustible resource." Greed, ignorance, politics, and federal mismanagement all contributed to deplete this fishery resource. Five species of salmon form the Pacific salmon fishery, although sockeye (red) salmon and pink salmon are the major commercial species. Adult salmon are caught along the Pacific coast as they return to spawn in freshwater streams. Young salmon fry spend the first part of their life in fresh water, but then move out into the Pacific Ocean where they grow to maturity in salt water.

Many Indian tribes along the coast depended on salmon as a staple food before the white man began fishing salmon commercially. Commercial exploitation in Alaska began about 1880 and increased at a slow rate (Figure 4.3). Most of the salmon taken commercially have been used for canned salmon. The peak year of commercial exploitation was 1936, when 8,500,000 cases of canned salmon were packed (a case is 48 pounds net weight). Since then the packs have declined to the point that the 1976 pack was below the 1900 pack. This increase and subsequent decline in the commercial catch also occurred in neighboring British Columbia but on a different time scale, and consequently the demise of the Alaska salmon fishery cannot be blamed on global ecological changes.

Throughout this time the amount of fishing gear used in the Alaska salmon fishery steadily increased. There were about 200 seine boats operating in 1910 and over 1,000 in the 1950s. There were about 1,000 gill net boats operating in 1909 and about 9,000 in 1955. This increase in gear (and hence in fishermen) was accompanied by a striking drop in catch per unit of gear. The average gill net boat caught about 15,000 salmon in 1908 but only 1,500 salmon in 1954, a 90 percent decrease. Thus more and more fishermen have been catching fewer and fewer salmon. This apparent violation of common sense becomes understandable in economic terms. The real price of a can of salmon has about tripled from 1910 to 1955, and this permitted the average fisherman to increase his income, at least until 1945 when it began to decline slightly.

The core of this overfishing problem and many others lies in the peculiar

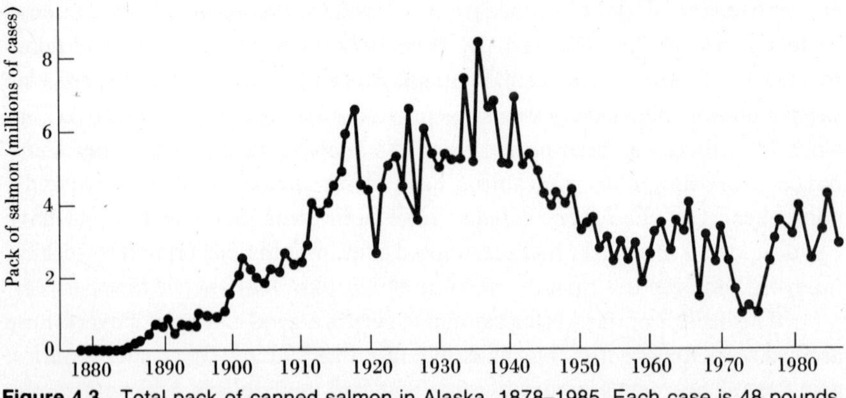

Figure 4.3 Total pack of canned salmon in Alaska, 1878–1985. Each case is 48 pounds (22 kilograms) net. (Data from *Fishery Statistics of the United States*)

common property status of the fishing resource. Salmon fishing in the past has been open to everyone who has the equipment, and the only way to limit the catch was to limit the efficiency of fishermen and gear. This situation reached the point where the salmon fishery was legally closed to fishing five or six days a week. Complex regulations rely heavily on law enforcement and put no premium at all on voluntary restraint of the fishermen or the cannery owners. There is, on the contrary, a positive economic incentive for each individual to catch as much as possible, and this is the root cause of overfishing. Salmon fishing now is strictly controlled by a license system but there are still too many boats licensed, and the fishery is closed more days than it is open.

Part of the reason for the overfishing of the Alaska salmon can be found in ecological ignorance. There was almost no information on the size of spawning populations for most of the major rivers of Alaska until the late 1940s. Population changes in salmon are still not understood, and the effects of fishing on stocks of salmon are unclear. Harvesting of salmon operates in the ocean on mixtures of stocks from different river systems and from different spawning areas within one system. One result is that less productive salmon stocks are overfished, and even driven extinct, while more productive stocks are not fully utilized. Part of the recent decline in the salmon catch may be caused by this kind of overfishing.

Both Alaska and Canada began programs in the 1970s to restore the former abundance of Pacific salmon. These programs are being aimed at the freshwater stages of the salmon life cycle on the assumption that if we can improve survival and growth in fresh water, the ocean will be able to support an increased stock. By a combination of improvements in natural streams, construction of artificial spawning channels with proper gravel and ideal water flow, and the use of hatcheries, fishery biologists hope to arrest the long-term decline of the Pacific salmon fishery.

These large-scale manipulations of salmon populations are now judged to be successful. Figure 4.3 shows that since 1975 the Alaska salmon fishery has recovered toward the pre-1940 levels. Salmon stocks in British Columbia are also increasing as a result of stringent restrictions on commercial and recreational fishing.

The history of the exploitation of whales has been a sad tale of overharvesting extending over 200 years. Many whale species have been driven to commercial extinction, the point at which costs exceed revenue and the whaling fleet moves elsewhere. Some species are near biological extinction as well.

The earliest commercial whaling operations were for the right whales *(Balaena glacialis)* in the North Atlantic. Right whales, so called because they were slow and easy to hunt and floated after being killed, supported the great arctic whaling industry from England, Scotland, and Holland in the 1700s and 1800s. By 1850 only a few right whales were left and this whaling fishery collapsed. Even now, after more than a century without any exploitation, the North Atlantic right whale is extremely rare.

The New England sperm whale fishery built up rapidly in the early 1800s and covered all the warmer oceans. Sperm whales *(Physeter macrocephalus)* were not as seriously depleted as right whales. The discovery of petroleum as a better source of oil for lamps destroyed the main markets for sperm whale products, and the industry declined for economic reasons, hastened by the Civil War (1860–1865), before sperm whales became too rare.

Early whaling was done from open boats with harpoons thrown by hand. Modern whaling dates from 1868 when a Norwegian, Svend Foyn, invented the harpoon gun and the explosive harpoon. With the harpoon gun whalers could go after the great blue and fin whales of the Antarctic. The greatest era of whaling thus began in the southern oceans in the early years of this century.

Whale catches in the Antarctic are shown in Figure 4.4. Whalers concentrated on three major species in succession—first blue whales *(Balaenoptera musculus),* then fin whales *(B. physalus)* and finally sei whales *(B. borealis).* Peak catches of blue whales were maintained for only a short period in the 1930s. Fin whale exploitation peaked from 1953 to 1962, and the stock collapsed during the next four years.

The International Whaling Commission was set up in 1946 in order to conserve whale stocks by regulating the whaling industry. Unfortunately, almost no scientific information was available in 1946, and the commission set limits on the total Antarctic catch that we know in retrospect were too high. The catch limit also did not distinguish species of whales, but set the allowable catch in blue whale units (1 blue whale = 2 fin whales = 6 sei whales). Blue whale stocks were thus driven down even farther through the 1950s and early 1960s and reached commercial extinction in 1963. Further restrictions on species and the allowable catch were gradually introduced in the early 1960s. By 1966 the quotas were in line with the allowable catch, and fin and sei whale stocks have stabilized at low levels. However, the Antarctic whale stocks have been depressed to such low population levels that the allowable catch is small. If all nations could agree to reduce whaling for 20 to 30 years to allow whale populations to increase to moderate abundance, we could maintain a whale fishery with a much larger allowable catch.

The public's awareness and horror at the collapse of the Antarctic whale populations have been all after the fact. Most people believe that we should conserve whale populations in the future, and the general message of the past history of whaling is that reliable scientific information on abundance, reproduction, and mortality is necessary for proper management. The key to managing any harvested population is to monitor the *unharvested* animals. If we knew in the 1950s what we know now, we could have arrested the decline of the Antarctic whale stocks for everyone's ultimate benefit.

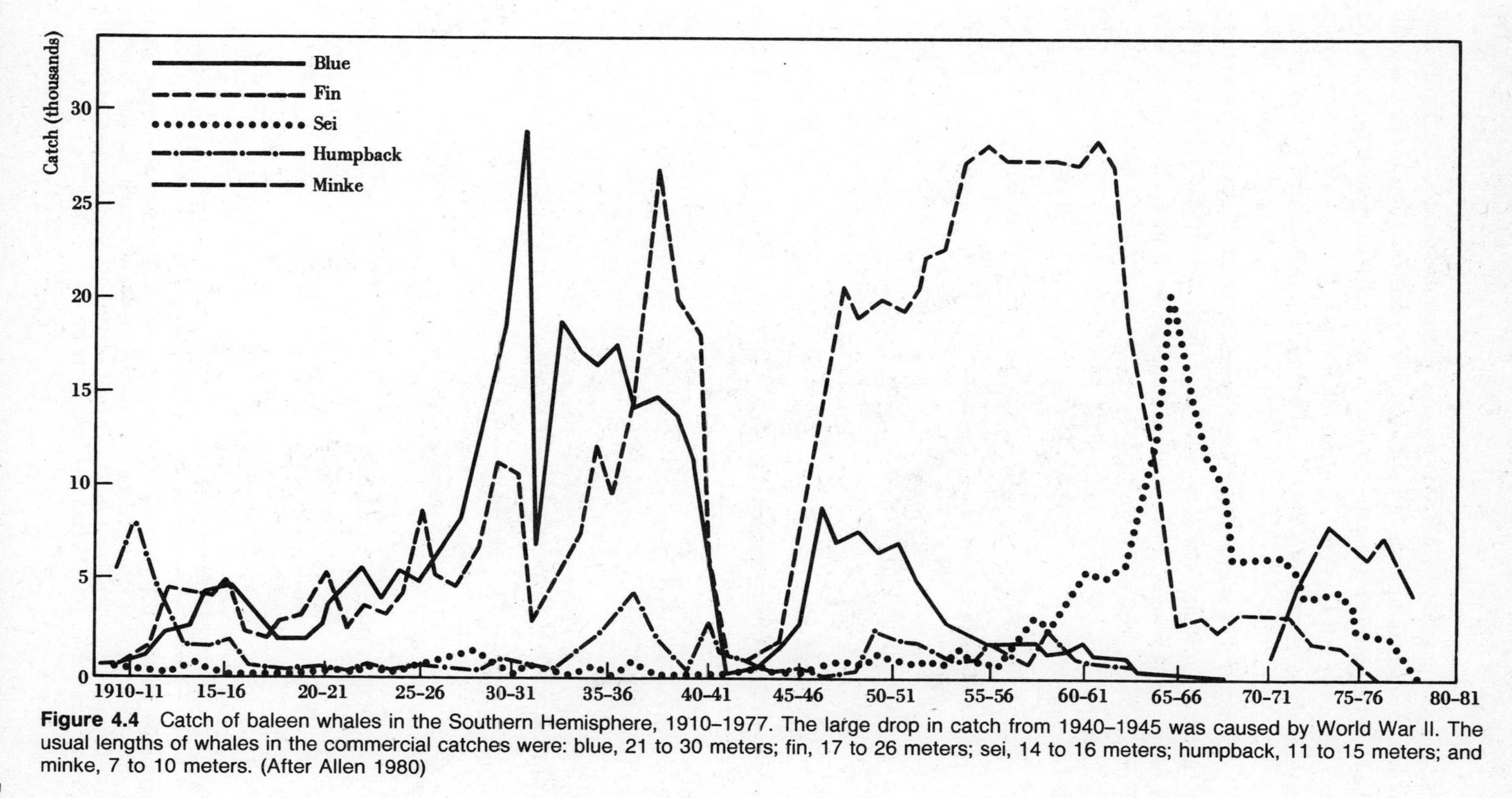

Figure 4.4 Catch of baleen whales in the Southern Hemisphere, 1910–1977. The large drop in catch from 1940–1945 was caused by World War II. The usual lengths of whales in the commercial catches were: blue, 21 to 30 meters; fin, 17 to 26 meters; sei, 14 to 16 meters; humpback, 11 to 15 meters; and minke, 7 to 10 meters. (After Allen 1980)

Large animals, which have very low reproductive rates, are highly susceptible to overexploitation. Predation by humans on large mammals may have been the cause of the extinction of many animals in prehistoric times. At the close of the last Ice Age (Pleistocene), some 10,000 to 12,000 years ago in North America, there was a massive extinction of large mammals. Over a short time 75 percent of the large mammals of North America became extinct. Why?

Extinctions of animals and plants have occurred throughout evolutionary time but typically on a slow time scale, with the orderly replacement of old forms by new ones. Pleistocene extinctions were unusual in being very

(a) *Megalonyx*, Jefferson's ground sloth

(b) *Mammut*, a mastodon

Figure 4.5 Two large herbivores that became extinct in North America at the end of the Pleistocene, about 11,000 years ago. (a) Jefferson's ground sloth was a woodland animal that has been found in fossil remains from central Alaska to Indiana. (b) The American mastodon was a browsing animal and has been found in over 1,000 localities as far south as Texas and Florida. (After Martin and Guilday 1967)

sudden, in striking only the large terrestrial herbivores and their predators, and in a lack of replacement. The net result is that we live in an impoverished world from which a set of large and strange animals has recently disappeared. In North America alone so many species disappeared at the end of the Ice Age that we can hardly believe what used to be here: ground sloths (Figure 4.5), giant armadillos, giant beavers, saber-tooth tigers, camels, mastodons (Figure 4.5), mammoths, several species of horses, giant deer, woodland musk-oxen, and other extinct forms less well known. This massive extinction was not paralleled by the extinction of small mammals on land or of small or large animals, such as the whales, in the oceans. Nor is there any evidence of large-scale extinctions in the plant kingdom. These extinctions could not have come from a global event like a large meteorite striking the earth because they did not occur at the same time on all the continents.

Two hypotheses have been suggested to explain these extinctions. The first is that *climatic changes* at the end of the Ice Age may have reduced the available habitat for these large animals and thereby eliminated them. Growing seasons became shorter and periods of food shortage may have become more frequent. The climate in North America became more seasonal and drier as the Ice Age ended, and grasslands expanded to cover areas that were formerly forest or woodland, but the new grasslands were simpler with fewer species of plants for large grazing animals to eat.

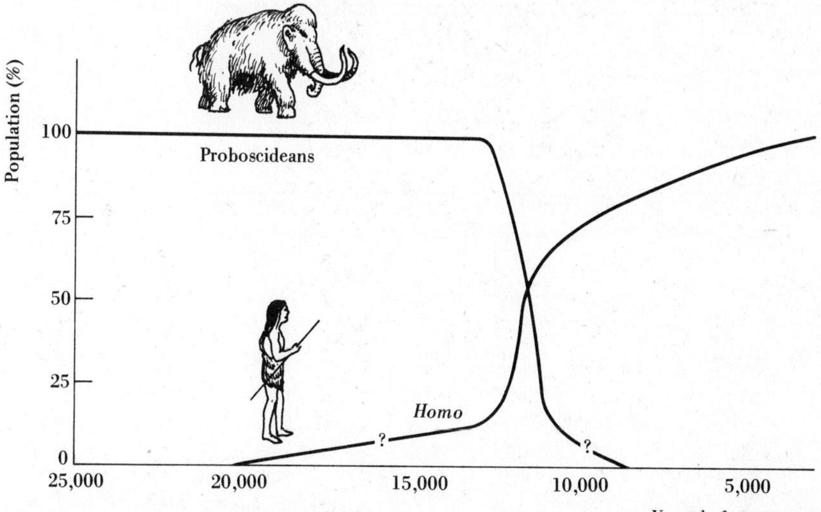

Figure 4.6 Time sequence of the extinction of the mammoths and mastodons in North America at the end of the last Ice Age and the accumulated incidence of human archaeological sites in North America. A good correlation like this is evidence in favor of the overkill hypothesis. (After Agenbroad 1984)

The second hypothesis is that prehistoric humans were responsible for these losses of large animals by *overharvesting.* Alfred Wallace first suggested this idea in 1911, but people found it hard to believe that humans who lacked any powerful technology could have such an enormous biological impact. However, more and more evidence is accumulating to suggest that this second hypothesis—the overkill hypothesis—could be correct.

The hypothesis of prehistoric overkill could be rejected if we could find cases of the survival of the extinct animals well after the coming of the early hunting tribes or cases of massive extinctions anywhere before humans arrived. In North America, where fossil data are relatively good, there is a close correlation between the extinctions and the immigration of early humans.

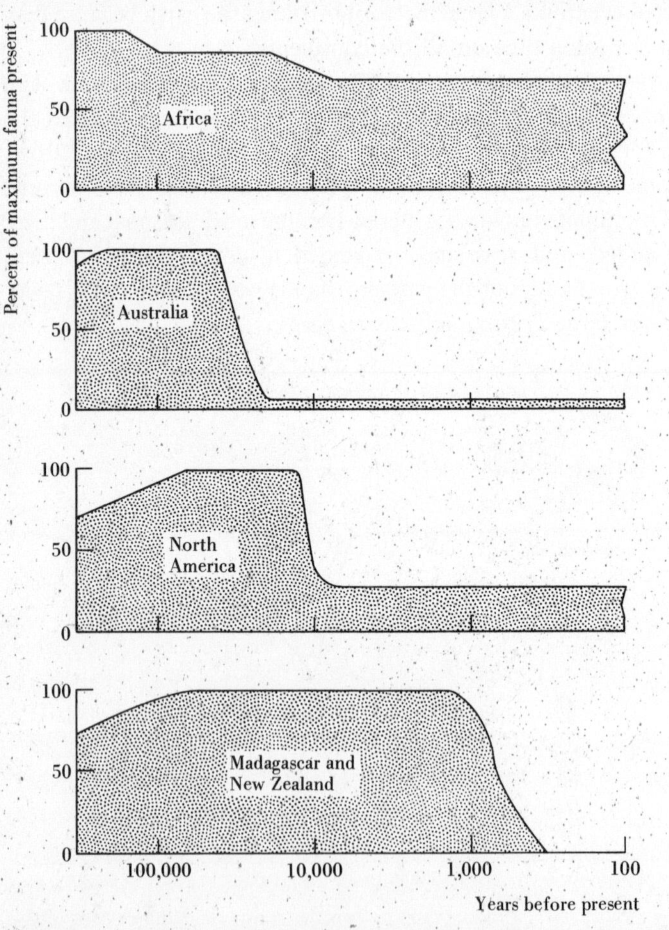

Figure 4.7 Global pattern of the recent extinction of large animals. Extinction was less severe in Africa compared with North America and Australia, but the reasons for this are not clear. In each case the major wave of late Pleistocene extinction does not occur until prehistoric hunters have arrived. (After Martin 1984)

Figure 4.8 Musk-oxen *(Ovibos moschatus)* herds have become much less common in the Canadian north during the last three centuries. (Photo courtesy of S. Groves)

Wooly mammoths disappeared in North America at the same time as humans colonized the New World (Figure 4.6). Since humans colonized the different continents at different times, the overkill hypothesis could be tested by seeing if large animal extinctions follow shortly after human colonization on all the continents. Figure 4.7 shows that the global patterns do in general coincide. In Australia, which humans colonized much earlier than North America, the extinctions occurred between 15,000 and 26,000 years ago. In Madagascar and New Zealand, the extinctions occurred only 800 to 900 years ago, the same time that hunters arrived on these large islands. The evidence of the coincidence of arrival of prehistoric humans and the demise of large mammals is too strong to dismiss. Although we consider ourselves, technological man, to be the main destroyers of species on the globe, the available data suggest that prehistoric hunters destroyed far more large mammals than we have.

There is one recent case of people with hunting-gathering technology greatly diminishing the numbers of a large mammal. Musk-oxen in the central Canadian arctic were formerly common both on the tundra and in the boreal forest (Figure 4.8). They were most common in a band of tundra lying just north of the tree line (Figure 4.9). Musk-oxen were first seen in 1689 by Europeans and were judged relatively abundant until about 1800. Their range began to contract and by 1860 musk-oxen were uncommon in the forest south of the tree line. Their decline accelerated in the late 1800s, and by the early 1900s this species was extinct south of the tree line. Only a few small, widely scattered populations have managed to survive south of the Arctic Circle during this century.

Musk-oxen were a critical resource for both Chipewyan Indians and Caribou Eskimos in the eighteenth century, and overhunting by these native hunters is believed to be the cause of the demise of these populations. Europeans seemed to contribute little to the destruction of musk-ox herds. Virtually no musk-oxen were killed by Europeans before 1890 and very few even after

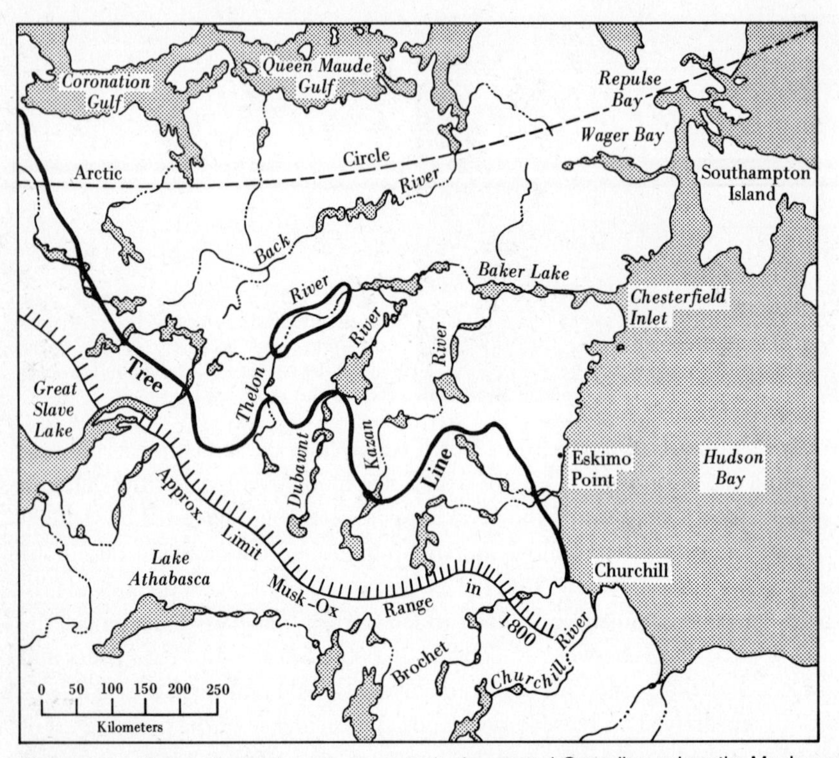

Figure 4.9 Historic limits of musk-ox range in the central Canadian subarctic. Musk-oxen were formerly common up to 200 kilometers south of the tree line (dotted line), but only a few small isolated herds still exist in this area of the Canadian subarctic, none south of the tree line. (After Burch 1977)

this date. Moreover, rifles supplied to the native peoples in the 1800s were not used for musk-ox hunting because these animals could be killed with traditional weapons. When musk-oxen are held at bay by dogs, hunters can approach with safety to within a few meters and kill them with spears or arrows. The fur trade in musk-ox hides may have created some additional hunting pressure during the late 1800s, but the Caribou Eskimos killed musk-oxen for meat, not hides, and they did not become actively involved in the fur trade until after 1920.

The suggestion from this historical analysis is that the native peoples of the central Canadian arctic overexploited musk-oxen populations and caused local extinctions over large areas without the use of guns. The effects of the demise of the musk-ox became evident during the famines of 1917 to 1921, when hundreds of Eskimos died from starvation. Caribou, their major food source, decreased for reasons unknown, and no musk-oxen were left to kill for food.

The human role in the extinction of many animals and plants has been a focal point for conservation groups during the last century. The destruction

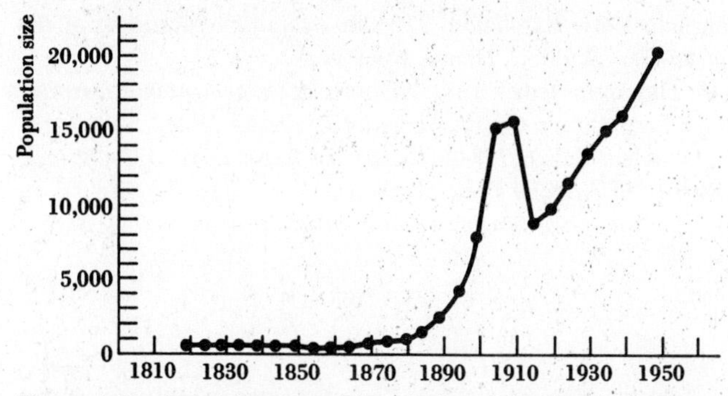

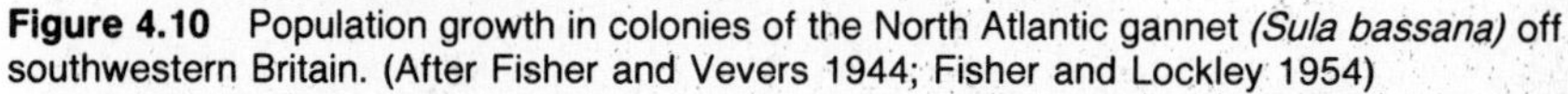

Figure 4.10 Population growth in colonies of the North Atlantic gannet *(Sula bassana)* off southwestern Britain. (After Fisher and Vevers 1944; Fisher and Lockley 1954)

caused by humans has fallen most severely on the larger animals, especially predators. Rare species and species on islands seem particularly vulnerable. In Scotland, for example, the brown bear *(Ursus arctos)* became extinct in the ninth or tenth century, and the wolf *(Canis lupus)* was driven extinct around 1700. The wildcat *(Felis silvestris)* disappeared around 1900, and the white-tailed sea eagle *(Haliaeetus albicilla)* was exterminated in the first half of this century. All these predators interfered with humans or preyed on domestic animals in Scotland, and hence they were persecuted and ultimately destroyed. This sad tale of extinction can be told for many animals in other parts of Europe and throughout the world. The twin forces of overhunting and habitat alterations have doomed many species to extinction.

Most species can recover from overharvesting if we recognize the problem quickly enough. The gannet *(Sula bassana)* is a sea bird that nests in large colonies on small islands. Gannets were exploited during the nineteenth century for their oil, for food, and for their feathers. Gannet eggs were also used for food. The result of this heavy exploitation was a great reduction of most gannet populations and the extinction of a few breeding colonies. By about 1880 protection was extended to colonies off southwestern Britain, and Figure 4.10 shows the subsequent recovery of these colonies, which are now thriving. Populations like the gannet's can spring back after being overexploited only if their habitat has not been destroyed. In the next chapter we shall look in more detail at the habitat changes that can complicate recovery from overharvesting.

FURTHER READING

Allen, K. R. 1980. *Conservation and Management of Whales.* University of Washington Press, Seattle.

Caughley, G. 1983. *The Deer Wars: The Story of Deer in New Zealand.* Heinemann, Auckland.

*Ehrlich, P., and A. Ehrlich. 1981. *Extinction: The Causes and Consequences of the Disappearance of Species.* Random House, New York.

Glantz, M. H., and J. D. Thompson (eds.). 1981. *Resource Management and Environmental Uncertainty: Lessons from Coastal Upwelling Fisheries.* Wiley, New York.

Martin, P. S., and R. G. Klein (eds.). 1984. *Quaternary Extinctions: A Prehistoric Revolution.* University of Arizona Press, Tucson.

May, R. M. (ed.). 1984. *Exploitation of Marine Communities.* Springer-Verlag, Berlin.

*Highly recommended.

chapter 5

Communities Can Rebound from Disturbances

Populations of plants and animals exist in a matrix of other populations, and we call all the animals and plants in a habitat a *community.* Communities are the most obvious ecological units because we see them every day—an oak forest, a grassland, a sagebrush desert—and one of the major jobs of ecologists is to try to understand how these communities work. Communities commonly contain thousands of species. How do all these species interact to produce the tapestry of nature? One way to see how something complex works is to take it apart. This approach can be applied to a biological community by studying the populations of each of the component species. But this is not the only way. Another way to see how something works is to perturb it—to cause some disturbance—and watch what happens. Humans have been disturbing natural communities on an increasing scale in recent years, and a casebook of oil spills, water pollution, logging, and pesticide treatments provides us with many examples.

Before we discuss some examples of perturbation we need to develop a general model of how communities can respond to disturbances. We can imagine a community as a billiard ball rolling on a topography set by the environment (Figure 5.1a). The ball comes to rest in a low spot in the topography, and we can call that resting position the natural or original community. If we disturb the system by hitting the ball slightly—the way we would perturb a community with a small oil spill—the ball rolls uphill a ways and then returns to the original position. We can use the jargon of physics to describe

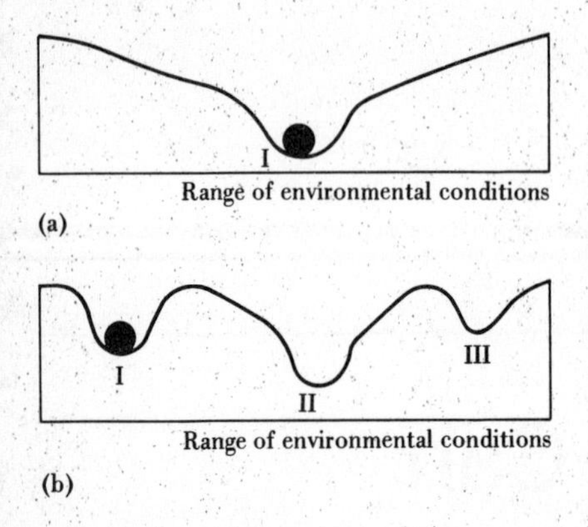

Figure 5.1 Local and global stability concepts. The community is represented as a black ball on a topographical surface which is a range of environmental conditions. In (a) the community is both locally and globally stable because after all perturbations it will return to configuration I. In (b) the community is locally stable, but if perturbed beyond a certain critical range, it will move to a new configuration (II or III). This community has *multiple stable points.*

this system as *locally stable,* and this is precisely what we mean when we say that communities can rebound from small disturbances. Next consider what happens if we clout the ball with maximal force, or provide a large perturbation. The ball may move quite a ways, even to the top of the hill and then, as before, roll back to its original position. This system is described as *globally stable* since no matter what you do to it, it returns to the original configuration. Biological communities are rarely globally stable, although some do approach this condition. Alternatively, we may find that a major force moves the ball to a new location on the topography (Figure 5.lb), that is, that the community does not return to its original configuration. The community may be locally stable, but if sufficiently disturbed, it will move to a new stable point. Communities of this type do not show global stability, but show several stable configurations. We will discuss these in the next chapter. In this chapter we will describe how communities respond to small disturbances and the mechanisms they have for maintaining local stability.

Two qualifications must be appended to the simple model shown in Figure 5.1. First, we must define what we mean when we say that a community changes its configuration. Ecologists usually mean that the species present in the community and their abundances change dramatically. We have all seen extreme examples of this type of change, for example, as a clear lake is polluted and changes to a green-water lake with scums of floating algae. Secondly, we must define the environmental conditions that form the perturbation and recognize that not all changes are equal. For example, temperate forests may be stable over a wide range of temperatures but sensitive to small changes in soil moisture levels. Or lakes may be sensitive to small amounts of phosphate pollution but unaffected by large amounts of nitrate pollution.

One job of the community ecologist is to find the limits of stability for natural communities and to see how they rebound from disturbances. To do this we will first consider the process of *succession.*

Our ideas about succession have been strongly influenced by the American plant ecologist F. E. Clements who in the 1920s developed a complete theory of plant succession. Clements suggested that there is a single end point to the development of all the communities in an area, so that they converge to a stable or *climax* community determined by climate. Clements' theory of succession is essentially that shown in Figure 5.1a—no matter where you begin, you end up at the same place, the climax community. An excellent illustration of this idea is provided by glacial moraine succession in southeastern Alaska.

During the past 200 years there has been a general retreat of glaciers in the Northern Hemisphere. As the glaciers retreat, they leave moraines (gravel ridges), whose age can be determined by the age of the new trees growing on them or, in the last 80 years, by direct observation. The most intensive work on moraine succession has been done at Glacier Bay in southeastern Alaska. Since about 1750 the glaciers there have retreated about 65 miles (105 kilometers), an extraordinary rate of retreat (Figure 5.2).

The pattern of succession in this area proceeds as follows. The exposed glacial till is colonized first by mosses, fireweed, alpine avens *(Dryas),* willows, and cottonwood. The willows begin as prostrate plants, but later grow into erect shrubs. Quickly the area is invaded by alder *(Alnus),* which eventually forms dense pure thickets up to 10 meters tall. This requires about 50 years. These alder stands are invaded by Sitka spruce *(Picea sitchensis),* which, after

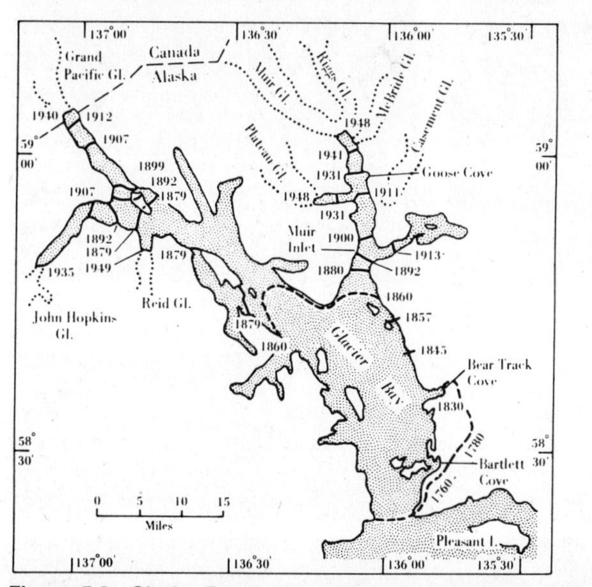

Figure 5.2 Glacier Bay fiord complex of southeastern Alaska showing the rate of ice recession since 1760. (After Crocker and Major 1955)

(a)

(b)

(c)

(d)

(e)

(f)

(g)

(h)

another 120 years, forms a dense forest (Figure 5.3). Western hemlock and mountain hemlock *(Tsuga mertensiana)* invade the spruce stands, and after another 80 years the situation has stabilized with a climax spruce-hemlock forest. This forest, however, remains only on well-drained slopes. In areas of poor drainage the floor of the spruce-hemlock forest is invaded by *Sphagnum* mosses, which hold large amounts of water and acidify the soil greatly. With the spread of conditions associated with *Sphagnum,* the trees die out because the soil is waterlogged and too deficient in oxygen for tree roots, and the area becomes a *Sphagnum* bog, or *muskeg.* The climax vegetation then seems to be muskeg on poorly drained areas and spruce-hemlock forest on well-drained areas.

The bare soil exposed as the glacier retreats is quite basic, with a pH of 8.0 to 8.4 because of the carbonates contained in the parent rocks. The soil pH falls rapidly with the invasion of vegetation, and the rate of change depends on the vegetation type. In contrast, there is almost no change in the pH from leaching of bare soil by precipitation. The most striking change is caused by alder, which reduces the pH from 8.0 to 5.0 in 30 to 50 years. The leaves of alder are slightly acid, and as they decompose they become more acid. As spruce begins to take over from alder, the pH stabilizes at about 5.0 and it does not change in the next 150 years.

The organic carbon and total nitrogen concentrations in the soil also show marked changes with time. One of the characteristic features of the bare soil is low nitrogen content. Almost all the pioneer species begin the succession

Figure 5.3 Plant succession as the ice retreats at Glacier Bay, southeastern Alaska. (a) Muir cabin site in the 1890s. The view is northwestward with the Muir Glacier front lying across the head of Muir Inlet of that time. Surface age is about 10 years. (Winter and Pond photo) (b) Close-up of fruiting, disk-shaped mat of *Dryas drummondii* about 1.5 meters in diameter. The white feathery fruit heads are ready for dispersal this dry day. This plant fixes nitrogen gas in its root nodules with the aid of symbiotic star-molds (Actinomycetes). Nearby are scattered willows and cottonwood saplings. Surface age is 15 years. (c) View north, north of Nunatak Knob, with continuous mat of fruiting *Dryas drummondii.* Muir Glacier has receded out of sight to the left, leaving an ocean fiord 300 meters deep. Scattered willows and cottonwoods became established before or simultaneously with the *Dryas.* There are two small thickets of alder in the center background and larger thickets on upper slopes of mountain on the right. Surface age is about 20 years. The photo was taken August 8, 1968. (d) Close-up of *Dryas drummondii* fruiting stalks opened and unfurling. The photo was taken at Muir Inlet, August 10, 1968. (James Taylor photo) (e) View northwest across Muir cabin site in the same area as (a) but 50 years later. Cottonwood and Sitka spruce tree crowns emerge from the alder-willow thicket 1 kilometer south of the Muir cabin site. Surface age is about 60 years. The photo was taken August 21, 1941. (f) Inside alder-willow thicket 400 meters south from Muir cabin site. Red elder shrub appears in the center and right foreground, with Sitka spruce in the center background. Surface age is about 60 years. The photo was taken August 21, 1941. (g) View southwest at Bartlett Cove, near the mouth of Glacier Bay, from which the glacier receded from A.D. 1750 to 1800. Sitka spruce forest, 30 meters tall, rises from the terminal moraine. The people on the beach provide a scale. Surface age is about 200 years. (h) Interior of first generation Sitka spruce forest with red squirrel midden of spruce cone-scales. Adjacent soil and stumps are covered with a *Hylocomium splendens* moss blanket. Glacial till surface age is about 200 years. The photo was taken at Bartlett Cove, August 14, 1968. (Photos courtesy of D. B. Lawrence)

with poor growth and yellow leaves caused by inadequate nitrogen supply. The exceptions to this are *Dryas* and alder; these species are able to fix atmospheric nitrogen in root nodules.[1] The rapid increase in soil nitrogen in the alder stage is caused by the presence of nitrogen-fixing microorganisms in nodules on the alder roots. Spruce trees have no such adaptation; consequently, the soil nitrogen level falls when alders are eliminated. The spruce forest develops by using the capital of nitrogen accumulated by the alder.

Plant succession is driven by the availability of limiting resources, most often *light* and *nitrogen.* Plant succession on glacial gravel begins with excess light and a shortage of nitrogen and ends with a shortage of light and adequate nitrogen. The important point to notice is the reciprocal interrelations of the vegetation and the soil. The pioneer plants that come first during succession alter the soil properties, these alterations permit new species to grow, and the new species in turn alter the environment in different ways, bringing about succession. The classical theory of Clements provides an adequate description of glacial moraine succession.

Not all succession follows such a smooth sequence as the glacial moraine succession just described. Consider a second example from farmland. Abandoned farmland goes through a succession back to native vegetation. When upland farm fields are abandoned in the Piedmont area of North Carolina, a succession of plant species colonizes the area.

Years after last cultivation	Dominant plant	Other common species
0 (fall)	Crabgrass	
1	Horseweed	Ragweed
2	Aster	Ragweed
3	Broomsedge	
5–15	Shortleaf pine	(Loblolly pine)
50–150	Oaks (hardwoods)	(Hickories)

However, early succession in Piedmont old fields is governed more by competition than by facilitation between plants. The early pioneers do *not* make the environment more suitable for later species, and the later species achieve dominance in spite of the changes caused by the early species rather than because of them. If seeds were available, broomsedge could colonize an abandoned field immediately rather than following horseweed and aster. Asters are slow to invade because horseweeds shade them and decaying

[1]Nitrogen fixation is the process by which organisms (often bacteria or fungi) convert atmospheric nitrogen (which most plants cannot use) into organic compounds containing nitrogen (which plants can use).

horseweed roots stunt aster growth. Broomsedge replaces aster because it is more efficient at taking up soil water and nutrients.

After the early succession by herbs and grasses, abandoned farmland of the Piedmont of North Carolina is invaded in great numbers by shortleaf pine *(Pinus echinata)*. Pine seeds can germinate only on mineral soil and are able to become established only when there is little root competition. The networks of pine roots in the soil become closed quickly, and the accumulation of litter under the pines as well as the shading causes the old-field herbs to die out. The density of pines is high, but falls rapidly after 15 years as the pines lose their dominance to hardwoods such as oaks *(Quercus)* and hickories *(Carya)* (Figure 5.4). Oak seedlings first appear after 20 years, when enough litter has been accumulated to protect the acorns from desiccation and the soil is able to retain more moisture. Hardwood seedlings persist in the understory because they develop a root system deep enough to exploit soil water and they are able to grow well in partial shade. Reproduction of shortleaf pine is almost completely lacking after about 20 years, because there is no bare soil for seed germination, and pine seedlings cannot live in the shade. After approximately 50 years several species of oaks become important trees in the understory, and the hardwoods gradually fill in the community.

Thus shortleaf pine is independent of the early succession in that it requires only bare soil for germination. The elimination of all herbaceous species from the early succession would not affect colonization by pines. Oaks and other hardwoods, by contrast, depend on the soil changes caused by pine litter, so that oak seedlings could not become established without the environ-

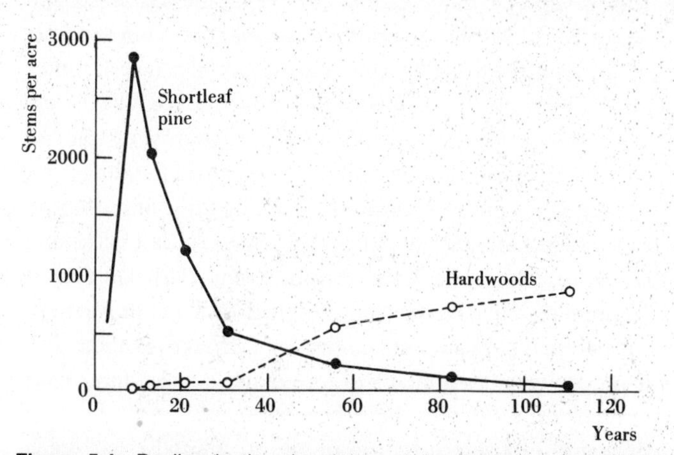

Figure 5.4 Decline in the abundance of shortleaf pine and increase in the density of hardwood tree seedlings during succession on abandoned farmland in the Piedmont area of North Carolina. (After Billings 1938)

mental changes produced by pines. Succession from pines to oaks can be interrupted by fire, which kills the oaks but not the pines. One trick pine trees use to prevent oaks from taking over is to produce foliage and litter that is highly flammable. In some areas of southern United States the occurrence of frequent natural fires results in a forest dominated by pines.

Forest succession in this and many other cases is governed by *germination requirements* and *shade tolerance.* One broad strategy for trees is to germinate in bare soil, grow rapidly in sunlight to large size, and scatter many small seeds widely. This is a strategy used by pioneer and fugitive species (see Chapter 3). A second strategy is to germinate in litter, grow slowly in shade, and produce fewer but larger seeds. To exist in a climax forest, a tree must be able to reproduce in its own shade, and thus trees of the climax forest tend to use some variant of the second strategy.

The importance of forest succession is clear when we consider the effects of logging. Logging is a form of disturbance and after logging is completed on an area, the community rebounds under the constraints imposed by succession. Succession can work to the forest industry's advantage or disadvantage depending on the types of trees it desires. In the Pacific Northwest, Douglas fir is a favored tree for timber, but when an area is clear-cut, alders seed in quickly and grow rapidly to overshadow Douglas fir seedlings. Unless foresters are prepared to wait 100 years for Douglas fir to replace the alders naturally, they must actively suppress alder succession and accelerate Douglas fir regeneration. This can be done by killing alders mechanically or chemically and then planting Douglas fir seeds or seedlings. Succession thus dictates how forest communities rebound from logging disturbances.

Some species that we find desirable cannot survive in climax forests and to preserve these species we must allow periodic disturbances. One example is the giant sequoia that requires fire to germinate successfully (see Chapter 3). Another spectacular example is the redwoods of California. Redwood trees *(Sequoia sempervirens)* in the north coast region of California grow on alluvial river flats and are subject to two kinds of disturbances—fire and flooding. Fire kills the seedlings of Douglas fir, grand fir, and bay trees; flooding and siltation kills trees of Douglas fir, grand fir, and tan oak. Only redwoods can withstand the normal disturbances in these river flats. By protecting redwood forests from fire and flooding humans have inadvertently signed their death warrant. These problems in our national parks are not always immediately evident. The vigor of a 1000-year-old redwood is difficult to assess and deaths may occur only tens of years after fire and flooding have been stopped.

White-tailed deer and moose are two examples of animals that reach peak abundance in successional forests and become rare in climax forest. Hunters are thus predestined to complain that "moose are becoming scarcer" as succession proceeds and the broad-leaved trees and shrubs that moose eat are reduced in numbers.

A good example of how a community can recover from a major disturbance was provided by the elimination of the American chestnut *(Castanea dentata)* from the eastern deciduous forests of North America. Chestnuts made up more than 40 percent of the trees in many climax deciduous forests, but they have been completely eliminated by the fungal disease chestnut blight. Chestnut blight was introduced to the New York City area around 1900 and proved lethal to all its hosts. By 1950 scarcely a single large chestnut tree was left living in North America, although the species survives as stumps that sprout. However, the effect of this removal on the other species in the community was slight. The eastern deciduous forests are rich in tree species, and oaks, beech, hickories, and red maple increased slightly in abundance and filled in the empty spaces left by the death of the chestnuts. No major changes in the animal and plant communities of the eastern deciduous forest have been ascribed to the loss of the chestnut.[2]

Pollution of our lakes and streams from sewage, industrial wastes, and agricultural runoff has resulted in ecological deterioration on a major scale. We need to know if it is possible to reverse these trends. In some cases complete recovery is possible once we eliminate the pollution. Consider one good example of how this can be achieved.

Lake Washington (Figure 5.5) is a large, clear, unproductive lake in Seattle, Washington. In the early phases of urban development Lake Washington was used for disposal of raw sewage, but this practice was stopped between 1926 and 1936. However, with additional population pressure, a number of sewage treatment plants were built between 1941 and 1959 and began to discharge treated sewage into the lake in increasing amounts. Sewage contains two major nutrients—nitrogen and phosphorus. By 1955 it was apparent that sewage was changing the clear-water lake, and a plan to divert sewage from the lake was voted into action. More and more sewage was diverted to the ocean after 1963, and almost all was diverted by March 1967.

What happened to the organisms in Lake Washington during this time? Since the diversion of sewage began in 1963, W. T. Edmondson has recorded the changes in the lake in detail. Some information can be obtained by looking at the sediments in the bottom of the lake. After sewage had been added to the lake, the sedimentation rate tripled to about 9 millimeters per year. The organic content of this sediment has progressively increased since the early 1900s, which suggests an accelerated rate of algal growth. The recent lake sediments also contain a greater amount of phosphorus. The composition of the diatom community in Lake Washington has also changed. The "shells" of diatoms are made of silica and are preserved well in sediments. A group of species of diatoms (the Araphidinae) varied in

[2]Bear populations may have suffered from the loss of the chestnut mast, but we have no good records.

Figure 5.5 An aerial view of Lake Washington, Seattle, Washington. The lake is 30 kilometers long and 5.6 kilometers wide at its widest point. The maximum depth is 62 meters. Water takes about 2.3 years to flush through the lake. Downtown Seattle is on the left and Puget Sound is visible in the upper part of the photo. (Photo courtesy of Skynet Aviation Inc., Seattle, Washington)

abundance in association with the sewage history and consequently can be used as *indicator species* of pollution in this lake.

Figure 5.6 shows the rapid drop in phosphorus in the surface waters and the closely associated drop in the amount of algae in the water since the diversion of sewage. Nitrogen content of the water has dropped little, which shows that phosphorus is limiting algal growth. The water of the lake has become noticeably clearer since the sewage diversion. Apparently the phosphorus tied up in the lake sediments is released back into the water column rather slowly.

Lake Washington did not remain unchanging after the sewage diversion was complete. In 1976 *Daphnia,* small crustaceans that feed on algae, became abundant. They had been held down by two factors before 1976. They were the main food of a predatory shrimp *Neomysis mercedis* that declined during the 1960s because of fish predation. They were also hampered by feeding on low quality algae *(Oscillatoria)* that clog their feeding apparatus. *Oscillatoria* declined after sewage was diverted in 1963. The net result is that since 1976 there are fewer algae and the water is even clearer in Lake Washington because *Daphnia* have cropped the algae to low numbers.

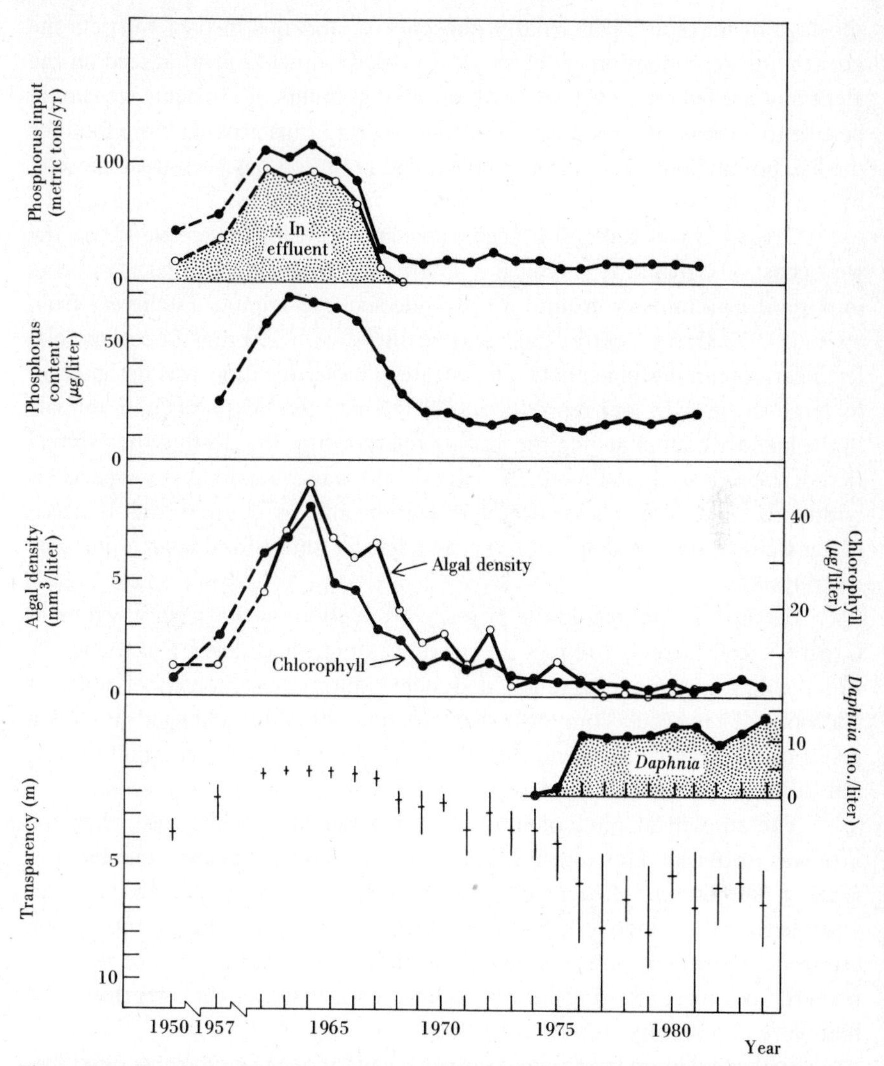

Figure 5.6 Recovery of Lake Washington, 1950–1984. Treated sewage flowed into the lake in increasing amounts during the 1950s. Sewage was diverted from the lake gradually from 1963 to 1968. The phosphorus content of the lake decreased rapidly after sewage was diverted. Algal density dropped in parallel with phosphorus levels because phosphate is the nutrient that limits algal growth in this freshwater lake. In 1976 the small crustacean *Daphnia* increased greatly in abundance and because it eats small algae, the algal abundance fell even more, so the lake water became clearer. Transparency is the depth to which a standard white plate can be seen in the water during midsummer. (Data courtesy of W. T. Edmondson)

The Lake Washington experiment is of considerable interest because it suggests that detrimental changes in lakes may be *stopped and reversed* if the input of nutrients can be stopped.

There is so much adverse publicity now about pollution that we tend to forget that some types of controlled "pollution" can be put to good use. The growth of microscopic algae (phytoplankton) in lakes is often restricted by a shortage of nutrients, particularly phosphorus, and this in turn restricts the growth and reproduction of the small animals (zooplankton) that feed on the algae and are fed on by fish. By adding small amounts of nutrients we should be able to increase the productivity of the lake and thus provide more food for the fish populations. This form of controlled pollution has been used successfully in some cases to improve fish production. Let us look at one example.

Great Central Lake is a large, clear lake on Vancouver Island, on the west coast of Canada. It is typical of a large number of coastal lakes in being important as a nursery ground for juvenile sockeye salmon. Each year from 1970 to 1973 Great Central Lake was fertilized with 100 tons of commercial fertilizer containing phosphate and nitrate. The fertilization was designed to increase the growth and reproduction of the microscopic plants and animals in the lake without changing the feeding relationships that lead to the production of young sockeye salmon. The experiment was evaluated by comparisons within the lake before and after fertilization and by comparisons between Great Central Lake and Sproat Lake, an adjacent unfertilized lake in the same river system.

Fertilization increased the primary production[3] of the surface waters of Great Central Lake to 10 times the previous, unfertilized level. For the whole water column primary production doubled during the months of nutrient additions. The species composition of the phytoplankton changed little as a result of this fertilization, and thus the integrity of the plant community was not disturbed.

The amount of phytoplankton in the water did not increase when the lake was fertilized. This puzzling response was due to increased zooplankton grazing. The amount of zooplankton increased over 10 times after the fertilizer treatment, and the increased zooplankton populations kept the phytoplankton cropped to low densities. No changes in the species composition of the zooplankton occurred after the addition of nutrients, and thus the integrity of the herbivore community was not disturbed.

Sockeye salmon fry hatch in the spring and move from streams into lakes to feed. These fry must stay in Great Central Lake for one or two years. Sockeye fry feed on zooplankton and thus should profit from the fertilization experiment. The average weight of juvenile salmon in their first year is shown in Figure 5.7 for 1969 (before fertilization) and 1970 (after). Juvenile salmon did, indeed, grow faster after the lake had been fertilized. Yearling salmon leaving the lake to enter the ocean averaged 72 millimeters in length in April 1970, but had increased to an average 79 millimeters in April 1971.

Survival of salmon at sea is partly related to their size when they leave fresh water. Most sockeye salmon return to Great Central Lake to spawn as

[3]Primary production is the amount of material made by green plants through their growth and reproduction; secondary production is the production of animals feeding on green plants.

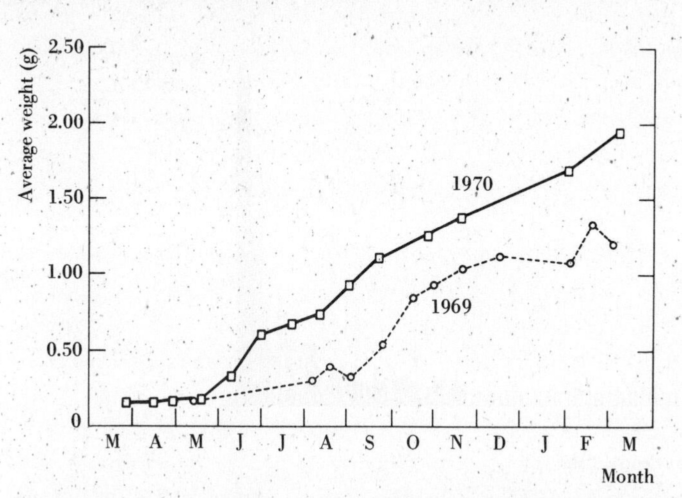

Figure 5.7 Average weight of sockeye salmon during their first year of life in Great Central Lake for 1969 (before fertilization) and 1970 (after fertilization). Juvenile salmon grew about 30 percent larger in 1970. (After Barraclough and Robinson 1972)

4-year-olds (71 percent) and 5-year-olds (27 percent), and so it was not until 1973 that any results of the fertilization could be observed. Figure 5.8 shows the tremendous increase in the abundance of adult sockeye salmon returning to Great Central Lake since 1973. The commercial catch increased 10-fold since fertilization became effective. The increased production of adults was evident both in Great Central Lake and in unfertilized Sproat Lake.

	Number of progeny returning per spawner	
	Before fertilization	**After fertilization**
Great Central Lake	1.7	7.4
Sproat Lake (control)	2.2	4.8

Because salmon home accurately to their natal stream, the spawning populations of these lakes cannot be mixing. Increased adult production from Sproat Lake may be an indirect effect of the experiment. Salmon from Sproat Lake probably had better survival in early marine life because of the Great Central Lake experiment, but it is not known why this has occurred. Juvenile salmon from both lakes go to sea at the same time and migrate together to the open ocean. If predation by other fish is important in early marine life, the large number of juvenile salmon may satiate the predators, and more juveniles may escape for both Great Central and Sproat salmon. If this is the correct explanation for the increase in Sproat Lake salmon, increases in salmon populations could be achieved by fertilizing one of a group of adjacent lakes in the same watershed.

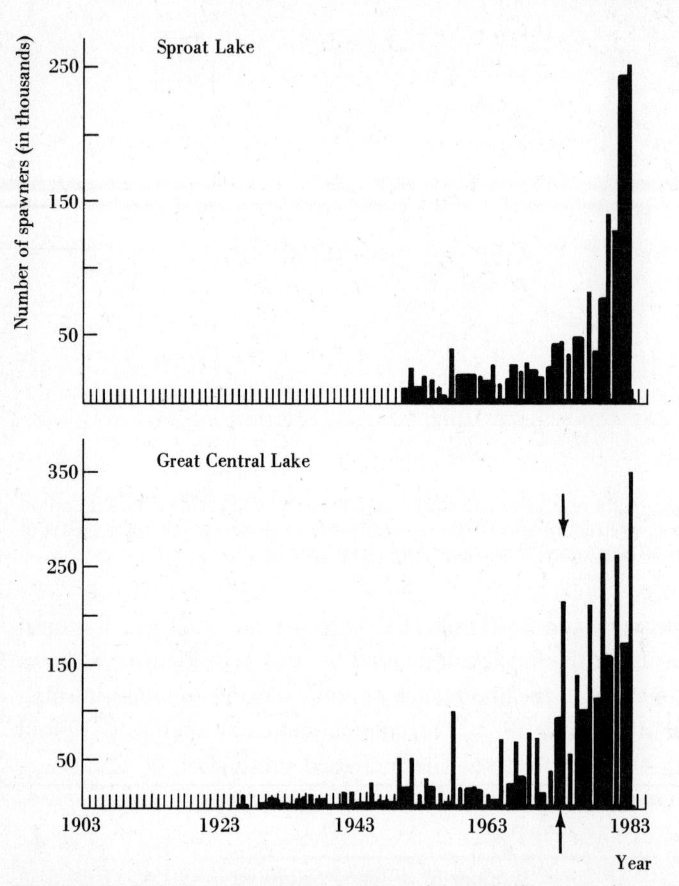

Figure 5.8 Estimated abundance (commercial catch plus spawning escapement) of Great Central Lake and Sproat Lake sockeye salmon, 1925–1983. The results of experimental fertilizations in Great Central Lake (1970–1973, 1977–1985) should be evident in sockeye runs beginning in 1973 (indicated by the arrows). Sockeye salmon abundance has increased at least 10-fold because of lake fertilization. (Data from K. D. Hyatt)

Nutrient additions to Great Central Lake were stopped in 1974 and resumed in 1977. One-year-old juvenile salmon in 1975 were the smallest observed since 1970. When nutrient additions were resumed in 1977, growth rates again improved in juvenile salmon, and adult salmon returns continued to increase to high levels. Because this fertilization program was so successful at Great Central Lake, it had been expanded to include 13 sockeye salmon nursery lakes by 1982, and it has already been judged a success in increasing salmon production.

The Great Central Lake experiment is a good example of how a detailed understanding of a community's response to nutrient additions can be used to gain practical advantages in the production of desirable species without altering the balance of other species in the community.

Oil spills are a highly visible form of pollution. Let us look at the ecological disturbances caused by a couple of oil spills that have been studied in some detail.

On the night of March 16, 1978, the supertanker *Amoco Cadiz* was wrecked 2.8 kilometers from the shore of Portsall, northern France. During the next two weeks most of its cargo of 223,000 tons of Iranian and Arabian crude oil spilled into the sea. About 360 kilometers of coastline were polluted from this disaster. In the first week after the wreck, heavy mortality of fish, crabs, sea urchins, clams, and other marine animals occurred. Between 15,000 and 20,000 seabirds died, mostly migratory birds nesting in Britain. The *Amoco Cadiz* spill also had long-term effects. Growth rates and reproductive rates in the commercial fish populations in the area were lower for at least two years after the spill than they had been before.

A freshwater marsh along the Mill River in Massachusetts was coated with an accidental spill of 3,800 liters of fuel oil on January 10, 1972. The oil was deposited on plants and soil throughout the marsh and persisted throughout the 1972 season. No detergents were used on this spill to disperse the oil.

Annual plants in the marsh were severely affected by the oil spill in 1972; all of them were either driven extinct or reduced in abundance. Many of the rare species disappeared completely after the oil spill and then began to reinvade the marsh in 1974 and 1975. Perennial plants were less affected immediately, and damage to some was seen only in the second year after the oil was deposited. The overall pattern was one of decline in the plant community during the two years after the spill, and a general recovery of vegetation through the next two seasons (Figure 5.9). But the warning is clear: Recovery

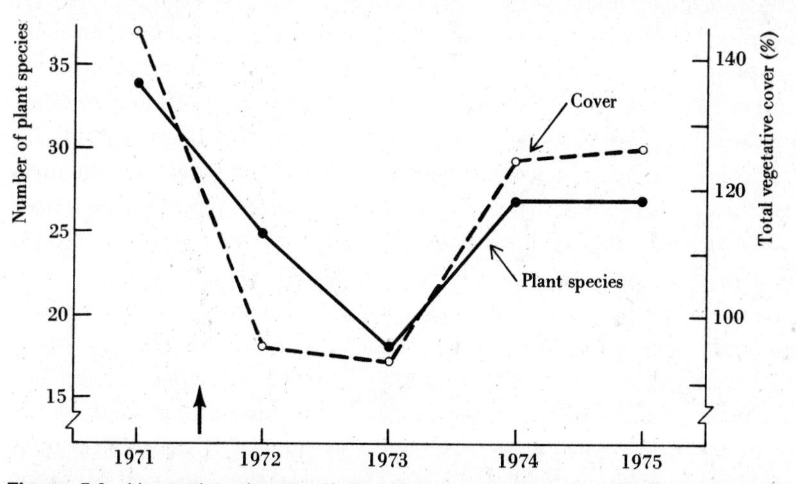

Figure 5.9 Vegetation changes in the high marsh zone of Hulbert's Pond, Mill River, Massachusetts, after a fuel oil spill on January 10, 1972 (arrow). (After Burk 1977)

even after four years was not complete, and the frequency of spills thus becomes crucial. The size of the area affected is also important because recolonization can occur only if seed sources are available.

Oil pollution in the ocean has many far-reaching effects on biological communities (Figure 5.10). All oils contain substances toxic to marine animals and plants. The most serious lethal effects on plants and animals are produced by the aromatic fractions of oil that are highly soluble in water. Different oils contain varying amounts of these soluble aromatic fractions and thus cause different degrees and types of damage.

Type of oil	Percent of soluble aromatics
Heavy crude oil	15
Medium crude oil	25
No. 2 fuel oil	30
Bunker C oil	5

Thus refined substances like No. 2 fuel oil can be expected to cause much more mortality of organisms than an equal amount of heavy crude oil.

In addition to immediate lethal effects, oil pollution damages organisms directly by coating and indirectly by altering habitat. All the chemical fractions of oil are important in coating. One of the most visible and pathetic effects of a spill is the kill of oil-coated birds. Oil kills seabirds by covering their feathers so that they lose insulation from cold and by preventing their feeding. Ingestion of oil from preening their feathers may cause physiological damage to birds and impair their ability to survive at sea. A few examples: An estimated 30,000 seabirds were killed by the *Torrey Canyon* spill on March 18, 1967. About 40,000 birds of 42 species were killed off the north coast of Holland in February 1969. Over 100,000 seabirds were killed off southwestern Alaska after two Japanese vessels sank leaving light diesel oil on April 25, 1970. It is impossible to judge the effect of these kills on seabird populations because the species all migrate to distant breeding places and insufficient data are available from many of these breeding sites to allow us to detect a drop in populations.

Another serious problem with oil pollution is that organisms accumulate hydrocarbons from oil in their tissues. If these hydrocarbons are not broken down, they may reach high concentrations. The taste of edible species, like oysters, may be affected, but more serious is the danger that some of the hydrocarbons accumulated may cause cancer in humans. Because of the dangers involved from tiny amounts of some chemicals, there is a real cause for concern, and a great deal of research is now being directed to this problem.

Some of the damage caused in early oil spills like the *Torrey Canyon* incident was produced by the detergents used in the clean-up campaign after

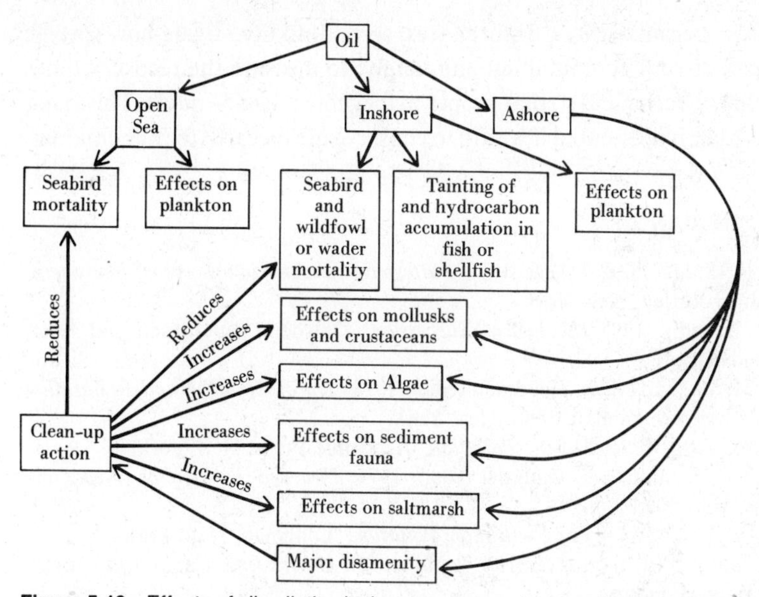

Figure 5.10 Effects of oil pollution in the ocean on marine organisms. (After Holdgate 1976)

the spill. The detergents succeeded admirably in causing the oil to break up and disappear sooner, but the detergents themselves were lethal to many organisms, more lethal in fact than the oil. This problem has been reduced in recent years by the development of physical means of removing oil and better chemical dispersants.

The recovery time of temperate zone marine communities after oil spills seems to be about 10 years. Clearly, the time scale of recovery through the process of succession will be different in tropical and in polar waters. Our best estimates so far come from the *Torrey Canyon* incident of 1967 off southern England in which lightly oiled rocks on a wave-battered coast took five to eight years to return to the normal community composition. Recovery can take much longer in bays and marshes with little wave action.

The ability of biological communities to recover after disturbance must not be perverted into a license to disturb natural communities. Ecologists can chart the recoveries of communities from small disturbances, but we should beware of extrapolating to much larger disturbances. A rocky coast may recover from a small oil spill in eight years, but it may not recover in a much longer time from a massive supertanker spill. Moreover, at present we can measure changes in only some of the common species in our communities, so we have only a crude ruler on which to measure biological disturbances. We should adopt the philosophy of treating natural communities with the same respect we treat our own bodies. We have all recovered from assorted diseases, but we know not to press our luck. We do not go about cultivating exposure to diseases. Our goal should be: first, to understand how communities of plants and animals operate without disturbances; second, to investigate how specific disturbances disrupt communities; and finally, to measure the recovery times from specific disturbances so that proper management can be made. This is not an easy task but it is essential if we are to conserve life on earth for our children.

FURTHER READING

Bormann, F. H., and G. E. Likens. 1979. *Pattern and Process in a Forested Ecosystem.* Springer-Verlag, New York.

Golley, F. B. (ed.). 1977. *Ecological Succession.* Dowden, Hutchinson and Ross, Stroudsburg, Pa.

Holdgate, M. W., and M. J. Woodman (eds.). 1978. *The Breakdown and Restoration of Ecosystems.* Plenum Press, New York.

Johnstone, R. (ed.). 1976. *Marine Pollution.* Academic Press, New York.

Kozlowski, T. T., and C. E. Ahlgren (eds.). 1974. *Fire and Ecosystems.* Academic Press, New York.

Mason, C. F. 1981. *Biology of Freshwater Pollution.* Longman, New York.

Smith, J. E. (ed.). 1968. *'Torrey Canyon' Pollution and Marine Life.* Cambridge University Press, London.

chapter 6

Communities Can Exist in Several Stable Configurations

A biological community can often rebound from a disturbance and return to its starting configuration. But if a disturbance is sufficiently drastic, or if the disturbing forces become permanent, the community may shift to a new configuration.

The configuration of a community is most clearly seen in its *food web.* The organisms in a community can be classified into three groups.

Producers—green plants
Consumers—animals
  Primary consumers—herbivores (eat green plants)
  Secondary consumers—carnivores (eat herbivores)
Decomposers—fungi and bacteria (break down dead plant and animal matter)

The major question in determining the food relations among members of a community is "Who eats whom?" The answers give us the food web of the community, such as the one shown in Figure 6.1. Food webs are organized by two major processes: "vertically" by *predation* and "horizontally" by *competition.* Thus, for the tundra of northern Alaska lemmings compete with one another for food or nesting sites, while tundra grasses and sedges compete for space in which to grow. Predation and competition always interact and what happens in the community is an overall summation of hundreds of

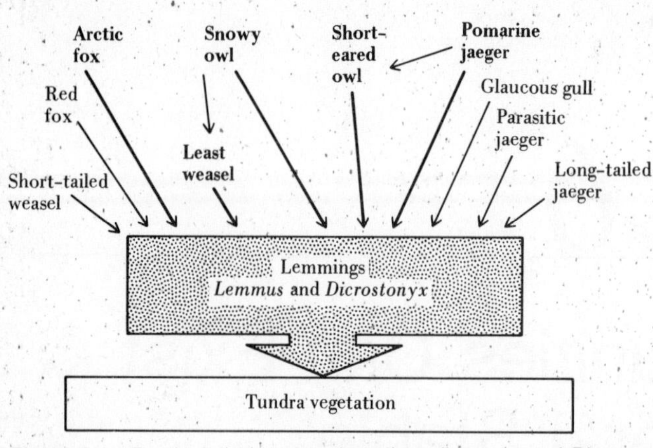

Figure 6.1 Food relations among lemmings *(Lemmus* and *Dicrostonyx)* and their predators in the Barrow region of northern Alaska. Names of the more important species among the various bird and mammal predators are shown in boldface letters. (After Pitelka et al. 1955)

species interacting through predation and competition. Perturbations to a community act through the linkages of the food web.

Food webs can grow very complex in any climatic or geographical region, even if only the major species are identified. Figure 6.2 shows the food web of the Antarctic Ocean and the central position of krill (large shrimp) and squid in this complex food web. Birds like the emperor penguin and animals like the leopard seal that both feed on squid can potentially compete for food. If strong pressure is exerted on any species in this web, it can reverberate throughout the web because of the tangle of feeding relationships.

The green plants, or producers, are usually the critical elements that fix the structure of the community. Nowhere is this clearer than in the oceans. The oceans can be subdivided into three types of areas that have very different food webs (Figure 6.3 on page 82). Over most of the ocean, the oceanic areas of Figure 6.3, there is a great shortage of nitrate and phosphate in the surface waters, and only very small plants called nannoplankton comprise the producers. These nannoplankton are eaten by small zooplankton (microzooplankton), and they in turn by small carnivorous crustaceans (macrozooplankton). But because of the small size of the plants at the start of the food web, up to five links must be passed before we reach the level of fish such as the tunas. Because energy is lost at each link in the food web, little of the energy fixed by the green plants reaches the level of the fish predators, and this food web is very inefficient as far as humans are concerned.

The food webs that occur over continental shelf areas and over regions of upwelling are shorter because they begin with larger sized algae. In areas of upwelling nutrient-rich waters from the ocean depths rise to the surface, and the high nutrient concentration supports the growth of large algal cells and colonies (macrophytoplankton). Fish such as Peruvian anchovy (planktivores)

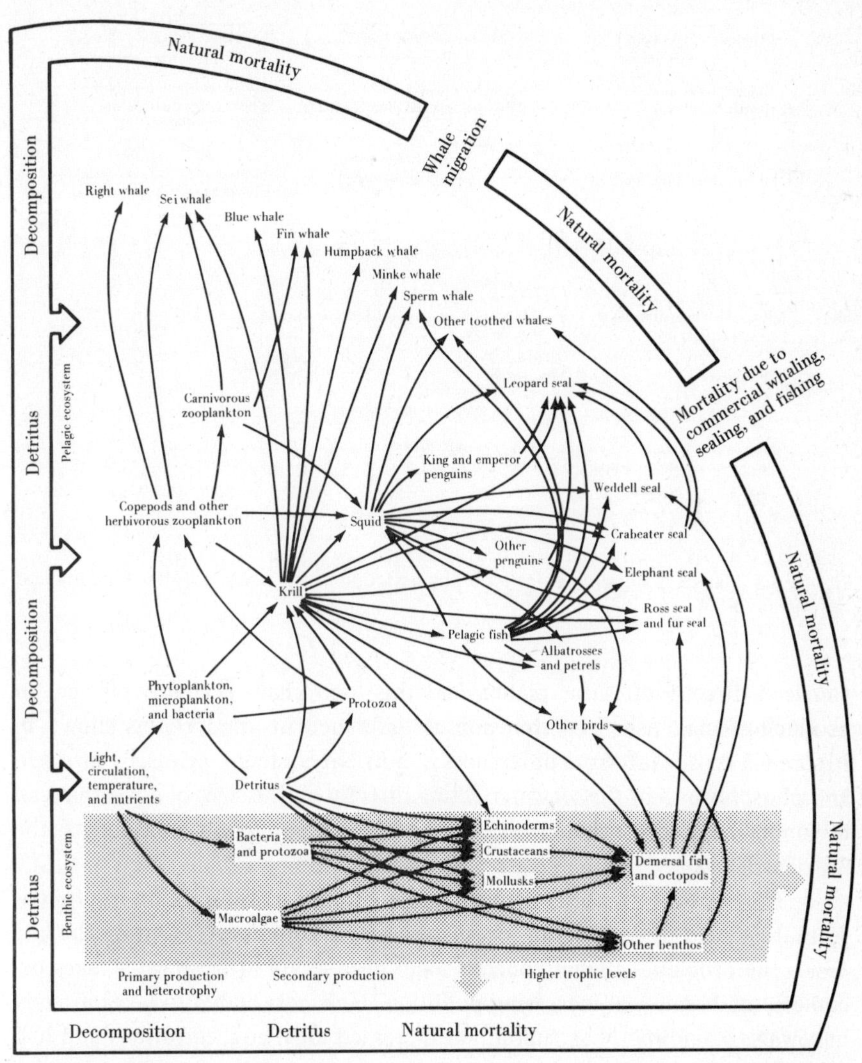

Figure 6.2 A food web for the Antarctic oceans. Primary production by phytoplankton and microplankton supports bacteria, protozoa, and zooplankton, including a krill population estimated at 500 to 750 million metric tons. Other key groups in the food web are copepods, carnivorous zooplankton such as chaetognaths, and some crustaceans, squid, and fish. Together with krill, these organisms directly or indirectly form the food base for the more conspicuous birds, seals, and whales. Benthic life depends on the rain of debris from above, and life in near-shore waters depends on primary production and detritus from seaweeds. Demersal fish living near the bottom feed on mollusks, crustaceans, and echinoderms such as sea urchins and starfish. Although the whales were once the principal consumers of krill, eating an estimated 190 million tons annually, their decline in numbers with whaling may mean that the available krill has been reapportioned, with birds and seals now taking the largest shares. (After Everson 1977)

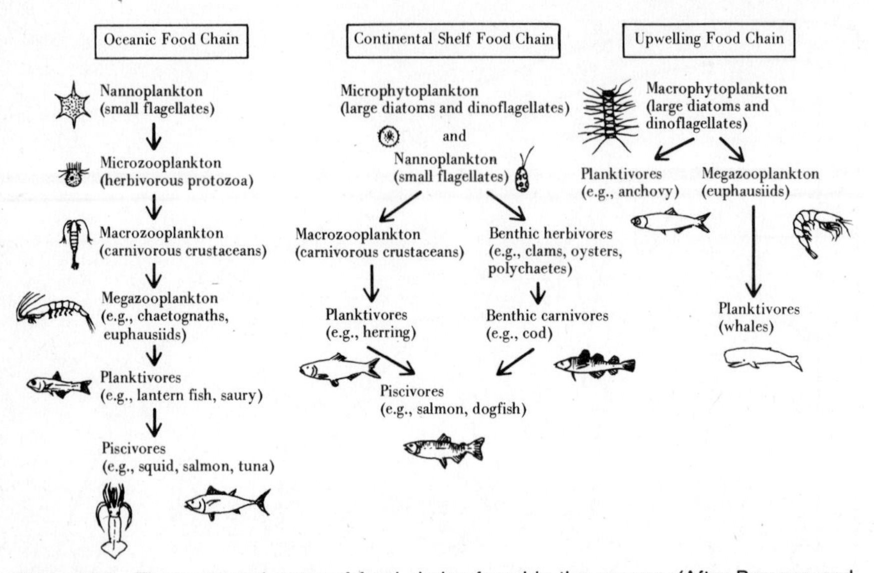

Figure 6.3 Three general types of food chains found in the oceans. (After Parsons and Takahashi 1973)

can feed directly on these plants, and this food chain is highly efficient in producing small fish. The tremendous differences in food chains shown in Figure 6.3 thus result from differences in nutrient content—primarily nitrogen and phosphorus—of the seawater acting directly on the sizes of algae that can be supported. Each of these food chains is a stable configuration as long as the nutrient levels are maintained.

The relation between nutrients and community organization is a fundamental element in considering disturbances to a community. During the late 1960s the problem of what controls algal production in freshwater lakes became acute because of increasing pollution. Nutrients added to lakes directly in sewage or indirectly as runoff had increased algal concentrations and had shifted many lakes from phytoplankton communities dominated by diatoms or green algae to those dominated by blue-green algae. This process is called *eutrophication.* Before anyone can control eutrophication in lakes, they have to decide which nutrients need to be controlled. Three major nutrients were suggested: nitrogen, phosphorus, and carbon. Phosphorus is now believed to be the limiting nutrient for phytoplankton production in the majority of lakes.

The Experimental Lakes area of northwestern Ontario has been used extensively for whole-lake experiments on nutrient addition. A series of well-designed experiments in these lakes has pinpointed the role of phosphorus in eutrophication. In one experiment, lake 227 was fertilized for five years with phosphate and nitrate, and phytoplankton levels increased 50 to 100 times over those of control lakes. To separate the effects of phosphate and nitrate, lake

226 was split in half with a curtain and fertilized with carbon and nitrogen in one half and with phosphorus, carbon, and nitrogen in the other. Within two months a highly visible algal bloom had developed in the basin to which phosphorus was added (Figure 6.4). All this experimental evidence is consistent with the hypothesis that phosphorus is the major limiting nutrient for phytoplankton in lakes.

The practical advice that has followed from these and other experiments is to control phosphorus input to lakes and rivers as a simple means of checking eutrophication. This advice has led soap manufacturers to develop

Figure 6.4 Lake 226 in the Experimental Lakes area of northwestern Ontario, showing the role of phosphorus in eutrophication. The far basin, fertilized with phosphorus, nitrogen, and carbon, is covered with an algal bloom of the blue-green alga *Anabaena spiroides*. The near basin, fertilized with nitrogen and carbon, showed no changes in algal abundance. The photo was taken September 4, 1973. (Photo courtesy of D. W. Schindler)

low phosphate detergents for household use. The permissible amount of phosphorus that can be added to a lake without causing serious pollution can be calculated so that planners can determine the desirability of human developments on a lake.

One of the changes that often accompanies eutrophication in freshwater lakes is that the blue-green algae tend to replace green algae. Blue-green algae are "nuisance algae" because they become extremely abundant when nutrients are plentiful and they form floating scums on highly polluted lakes. Blue-green algae become dominant in the phytoplankton for two reasons. They are not eaten very much by zooplankton or by fish, which prefer other algae, and they are more efficient than green algae at taking up carbon dioxide and phosphate from low concentrations. The phytoplankton community in many temperate freshwater lakes therefore seems to have two broad regions of equilibrium—one with low nutrient levels dominated by green algae and one with high nutrient levels dominated by blue-green algae.

Changing the nutrient input to green plants can have a profound impact on the type of community that develops in aquatic systems. Can we produce the same kind of effects by tinkering at the other end of the food web on the predatory animals in a community? The best work on this question has been done in freshwater lakes where new predators can be introduced.

The zooplankton community of many lakes in the temperate zone is dominated by large-sized species when fish are absent and by small-sized species when fish are present. This was first observed in Crystal Lake, Connecticut, after the introduction of a herringlike fish, the alewife *Alosa pseudoharengus* (Figure 6.5). How can we explain the observed shift in the zooplankton community from large to small species? Let us make three assumptions: (1) Zooplankton all compete for food (small algal cells) in the open water. (2) Larger zooplankton feed more efficiently on small algae than do smaller zooplankton. (3) Fish feed more easily and more effectively on larger zooplankton and so prefer to eat them.

Given these three assumptions, we can make two predictions: (1) When fish predators are absent or few in number, the small zooplankton will be eliminated by large forms because of competition for food. (2) When fish predators are common, they will eliminate the large zooplankton and allow the small zooplankton to become dominant. These two predictions seem to describe adequately the structure of the zooplankton community in many lakes as a joint outcome of the interaction between competition and predation.

The stresses of fishing can become a permanent feature affecting the biological communities in lakes and in the sea. At the same time pollution has increased so that aquatic communities are under stresses from two directions at once. What happens in lakes subjected to both heavy fishing and heavy pollution? Nowhere has this been more strikingly shown than in Lake Erie.

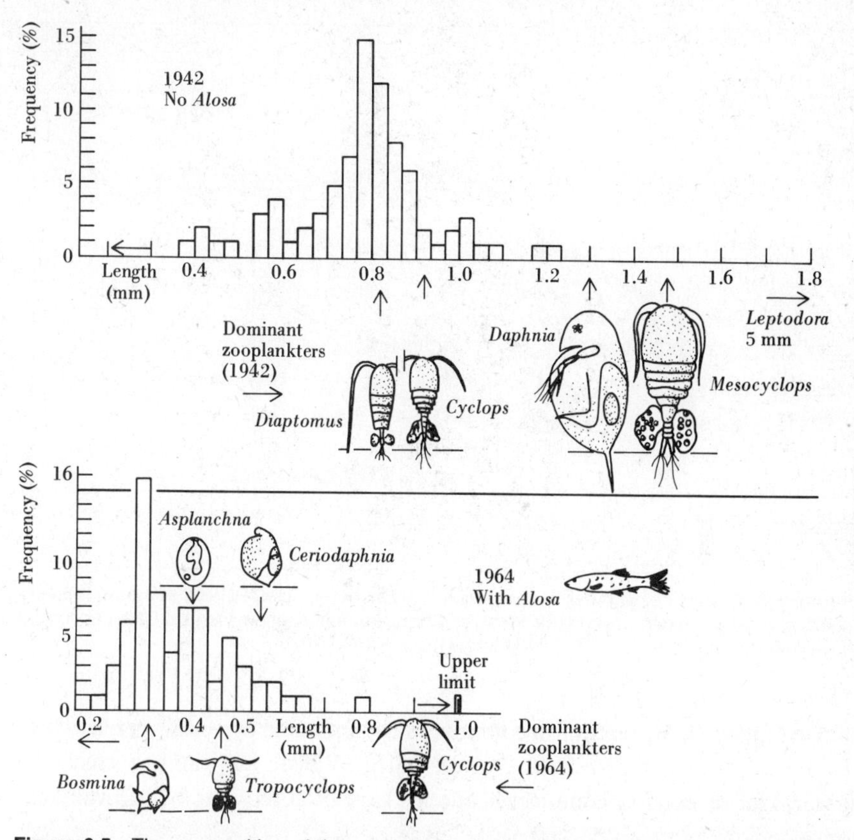

Figure 6.5 The composition of the crustacean zooplankton of Crystal Lake, Connecticut, before and after the introduction of the alewife *(Alosa pseudoharengus)*, a plankton-feeding fish. (After Brooks and Dodson 1965)

Lake Erie (Figure 6.6) has changed greatly in historical time. In 1800 the lake was bordered by large stands of forest and extensive marshes. Because of the vegetative cover, soil erosion was limited, runoff waters were clean, and aquatic vegetation flourished. By 1870 the area had changed to an agricultural one with woodlands being cleared and swamplands drained. Exposed soil from farmlands washed into the rivers and deposits of silt began to cover the spawning grounds of many fish. Nearly all the marshes and swamps were drained by 1910, which destroyed more spawning and nursery areas for fish.

In 1800 Lake Erie contained a great variety of fish. Large game fish like smallmouth and largemouth bass, muskellunge, northern pike, and channel catfish lived inshore. Lake herring, blue pike, lake whitefish, lake sturgeon, walleye, sauger, freshwater drum, and white bass were found in the open lake. Lake trout lived in the eastern end of the lake. Today the blue pike, sauger, and lake trout are gone, and few sturgeon, lake herring, whitefish, and muskel-

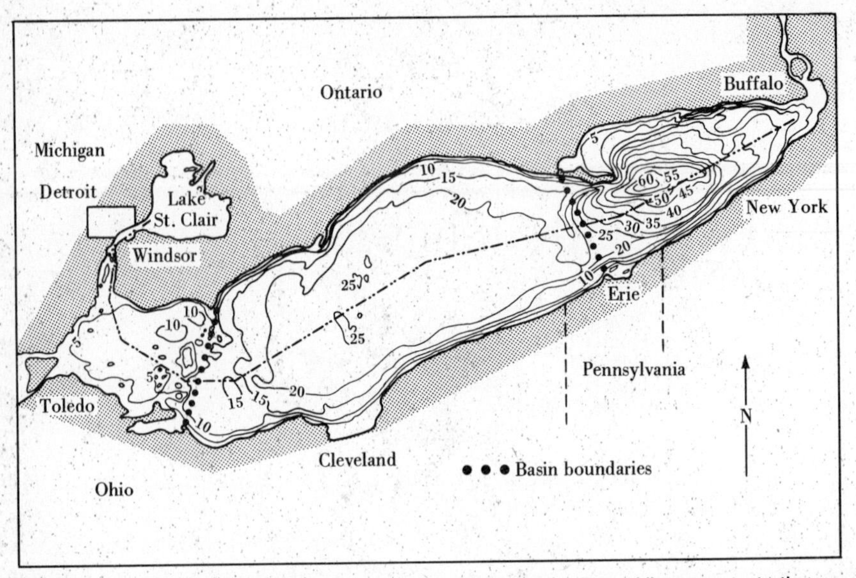

Figure 6.6 Lake Erie, with depth contours shown in meters. Dotted lines separate the very shallow Western Basin of the lake from the larger but still shallow Central Basin and from the deeper Eastern Basin. (After Regier and Hartman 1973)

lunge remain. The present fish community is dominated by yellow perch, white bass, channel catfish, freshwater drum, carp, goldfish, and rainbow smelt. In general more valuable commercial species have been replaced by less valuable ones.

The commercial catch of fish from Lake Erie over the past 150 years has exceeded that of the other four Great Lakes combined, and the fishery has been one of the major stresses on the lake communities. The commercial fishery began after the War of 1812 and developed rapidly after 1820, so that from 1800 to 1850 the catch grew about 20 percent per year (Figure 6.7). The economic value of the Lake Erie fishery was highest from 1881 to 1890.

The lake sturgeon was the first casualty of the Lake Erie fishery. Because the sturgeon was large (over 80 kilograms sometimes) and covered with heavy scales, it tore up the gill nets set for smaller species. Fishermen got heavier nets, caught sturgeon in large numbers, and then destroyed them by piling them on the beaches like cordwood, dousing them with oil, and burning them. In the 1860s an immigrant from Europe arrived with the knowledge of how to smoke sturgeon and make caviar from its eggs. By 1870 the sturgeon had become a valued fish, but it was already scarce from overfishing. After sturgeon numbers had decreased, the fishermen turned to lake trout in the 1880s and lake whitefish in the 1890s. Concern mounted among fishermen in the late 1800s because of the decline of these valuable fish stocks, and two solutions to the problem were advocated. One solution was to regulate the fishery stringently,

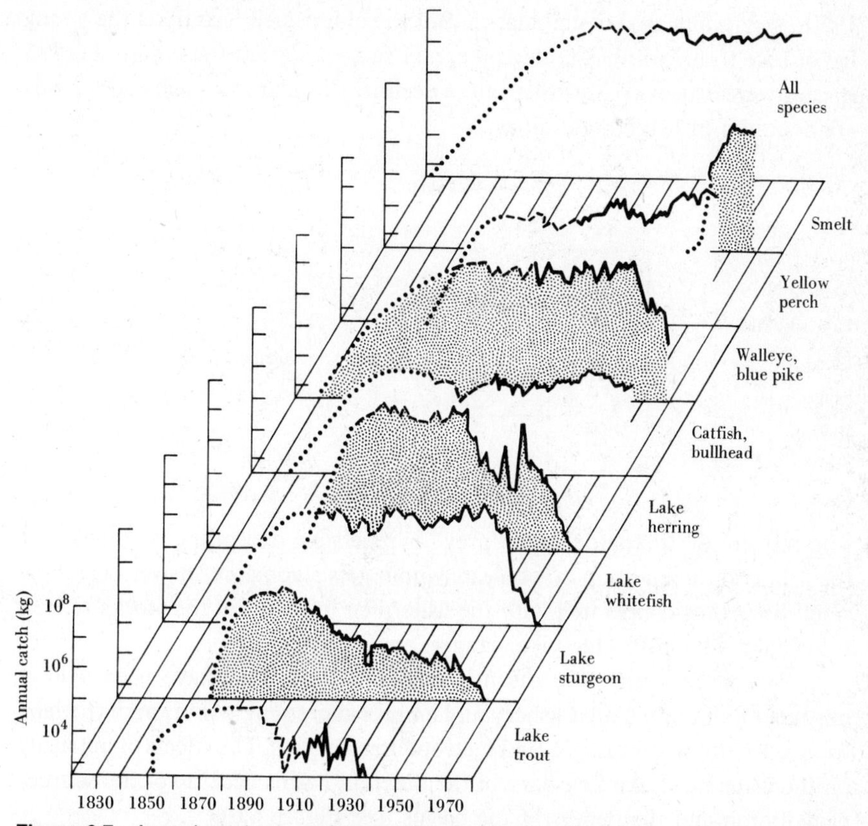

Figure 6.7 Annual catch of a number of fish species from Lake Erie, 1820–1971. Dotted lines are estimates based on historical accounts. The total catch of this fishery has been nearly constant since 1860. (From Regier and Hartman 1973)

and a few regulations were introduced. But the second solution was technological and appealed to free enterprisers. France in 1850 had developed a new technology—the use of fish hatcheries—with the aim of helping the natural reproduction of desirable fishes. The hatchery idea spread among North Americans and government agencies rushed to develop hatchery and fish stocking programs. Between 1867 and 1920 some 18 hatcheries were constructed on Lake Erie. All but one of these hatcheries have been quietly closed down after studies showed that virtually none of the fish fry stocked in Lake Erie had survived.

An alternative was to introduce new species of fishes into Lake Erie, and many introductions were made. Most failed but a few introduced fishes had catastrophic effects on the native fish population. Rainbow smelt were introduced into Lake Erie around 1931 and increased rapidly during the 1940s. Smelt live in the open waters of the lake and are eaten by lake trout, whitefish, blue pike, and sauger. As these species were reduced during the 1940s and

1950s, smelt increased in abundance. But the older smelt feed upon the young fry of lake trout, whitefish, blue pike, and sauger, so that these desirable fish species were driven even further into a decline. Circular food webs of this sort are common in fish communities.

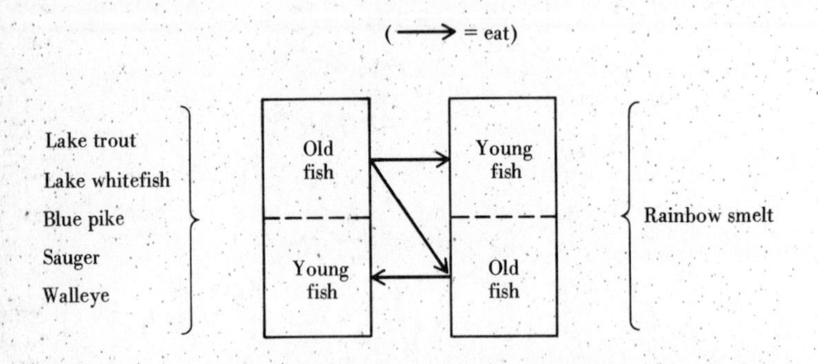

The concept of "predator" and "prey" is reversible depending on the size of the fish, and the question of who eats whom gets a complex answer. Rainbow smelt have thus cooperated with the fishermen in reducing the stocks of lake trout, lake whitefish, blue pike, sauger, and walleye.

The large changes in the fish stocks of Lake Erie have been a joint product of an unregulated fishery and an ever-increasing pollution of the lake from agricultural, industrial, and metropolitan sources. The effects of nutrients on the waters of Lake Erie vary in the different basins because of the sources of pollution and the depths of the basins (see Figure 6.6).

The Western Basin of Lake Erie is shallow and has the most valuable fish spawning grounds in the lake. It has been subject to excessive pollution from the cities of Detroit and Toledo. Blue-green algae have come to dominate the phytoplankton, and algal production has increased because of the nutrients added in sewage and industrial wastes. The large amounts of algae produced in the surface waters fall to the bottom of the shallow Western Basin and decompose, using up all the oxygen in the water. Mayfly larvae, once common in the bottom muds, have been replaced by bloodworms that can tolerate periods with no oxygen in the bottom waters.

The cold bottom waters of the Central and particularly the Eastern basins of Lake Erie are the summer sanctuaries of the valuable cold-water fishes—lake trout, herring, whitefish, and blue pike. As more and more nutrients have been added to Lake Erie, the Central Basin has become more productive, and its bottom waters have also lost all oxygen in the summer months. The net result is that suitable habitat for fish spawning, feeding, and resting has been more and more restricted, particularly for the native cold-water fishes.

The fish community of Lake Erie has thus moved to a new configuration under the impact of the following disturbances in order of importance.

most important	Uncontrolled overfishing
↓	Erosion and nutrient pollution
	Introduced species
	Stream destruction, dams, shoreline changes
least important	Toxic chemicals and pesticides

What is the prognosis for the fish community of Lake Erie if water pollution is reduced and nutrient inputs decline? Some recovery is possible, but the historical fish communities cannot be regained. Introduced species like the rainbow smelt would prevent the restoration of lake whitefish, blue pike, or lake herring. Further complications have been added by the recent introduction of Pacific salmon (coho and chinook) to Lake Erie in response to the demands of recreational fishermen. The recovery of the Lake Erie fish community rests directly on the need to provide continuous oxygen-rich bottom water in the Central Basin, and this can be achieved only by a reduction in nutrient inputs to the lake. Once this is achieved, cold-water fishes can again become abundant and a more desirable mix of species can be attained in Lake Erie.

We can try to generalize from the Lake Erie example to predict the community changes that develop from major stresses. Fish communities in freshwater lakes and continental shelf zones contain three broad groups of fish:

Large fishes: bottom dwelling, feed on fish, highly predatory, nonmigratory, low growth rates, older at maturity, populations stable in time, highly desirable species for fishermen (e.g., lake trout)

Medium-sized fishes: partly open-water dwelling, feed on small fish and plankton, moderate population fluctuation, desirable species for fishermen (e.g., walleye)

Small fishes: open-water dwelling, feed on plankton, high growth rates, younger at maturity, great population fluctuations, undesirable species for fishermen (e.g., rainbow smelt)

As a stress is applied to the fish community, the larger forms fade away gradually and the medium-sized forms collapse irregularly. The small fishes are the most resilient to stresses and tend to retain high but fluctuating abundances, so that the fish community becomes dominated by small plankton-feeding fish. If this general pattern of changes is recognized in lakes affected by human activity, we can design manipulations specifically to thwart the deterioration of the fish community. No one has done this yet for any lake, but as we gain more insight into the forces that alter communities, we should be able to move the fish community to the configuration that we think most desirable.

Many aquatic communities have changed greatly under the joint stresses of overfishing and pollution, but terrestrial communities have been altered even more drastically by human activity. Landscapes may change slowly, so the changes are not always obvious to us as they occur. Two examples will illustrate these ideas.

In terrestrial communities heavy grazing can produce profound changes in the type of plant community in the landscape. Some of these changes occurred so long ago that we scarcely recognize them. The heather moors of Scotland are an example (Figure 6.8). Much of the Highlands of Scotland is now covered with heather moors, treeless areas of short vegetation dominated by ling heather *(Calluna vulgaris),* and the beauty of the moors has been thoroughly enshrined in poetry and song so that we are almost conditioned to think of "Scotland" and "heather" together. It comes as a surprise then to find out that heather moors are *not* the natural climax vegetation of the Highlands, but are a community maintained by excessive grazing by sheep, rabbits, and deer!

Heather is a dwarf shrub with evergreen leaves. Several thousand years ago it was found principally as a shrub in open parts of pine, birch, and oak

Figure 6.8 Heather moor in eastern Scotland. The vegetation is predominately heather *(Calluna vulgaris).* The remains of tree roots and stumps show that this area was a mixed forest of Scot's pine and birch several hundred years ago. (Photo courtesy of N. Picozzi)

woodland on acid soils, and forest clearance was necessary before it could flourish. Forest destruction began in Scotland about 3000 B.C. Early humans burned and cut the woodlands to provide room for agriculture, and as human settlement increased, the destruction of forest land accelerated. At the same time the climate in Scotland became cooler and wetter, and grazing animals began to use the areas cleared of trees. The open moors provided good free-range grazing for sheep, and in Scotland stocks of hill sheep increased greatly in the 1700s and 1800s. Periodic fires were also started in the heather to maintain the production of new green shoots for sheep. The combination of burning and grazing by sheep, rabbits, and deer ensures that tree seedlings are destroyed and that the heather moor is maintained as a stable community. On many moors tree seed is now so scarce that even if burning and grazing were to cease, there would not be an immediate reversion to woodland. The plant community has moved to another stable configuration, one that humans consider the "natural vegetation" of the Highlands.

One way to see the effects of grazing on a plant community is to remove the grazing animals. A few spectacular natural experiments have occurred along these lines. The European rabbit *(Oryctolagus cuniculus)* has been in Britain for at least a thousand years and has been common for more than a hundred years as an agricultural pest. The impact of the rabbit's grazing was not clear until an epidemic of the virus disease myxomatosis in 1954 virtually eliminated rabbits from many areas in Britain. By 1960 rabbits were becoming resistant to the disease, and populations began to build up again. But for about six years we could see how the plant community changed in the absence of rabbit grazing.

After the disappearance of rabbits, grasses increased greatly in height and there was a spectacular increase in the abundance of flowers. Orchids previously known only from a few orchid leaves suddenly became evident in large flowering masses. Rabbit grazing thus seemed to produce floristic poverty. Another general effect of rabbit removal was the growth of woody plants. Brambles, gorse, heather, and tree seedlings of oak and ash began to appear in areas rabbits had grazed. Succession is thus different in the presence and in the absence of rabbits.

Rabbits, like any grazing animal, divide the world of plants into three broad categories: (1) preferred food plants, which decrease under grazing pressure, (2) edible but not preferred plants, (3) inedible plants, which increase under grazing pressure. The results of removing a grazing animal are determined by the competitive abilities of the individual plant species. Figure 6.9 shows this for a plot enclosed in a rabbit-proof fence in 1936. Two plants compete for dominance in the absence of rabbit grazing. *Festuca ovina* (sheep's fescue grass), which dominated the grassland under rabbit grazing, was re-

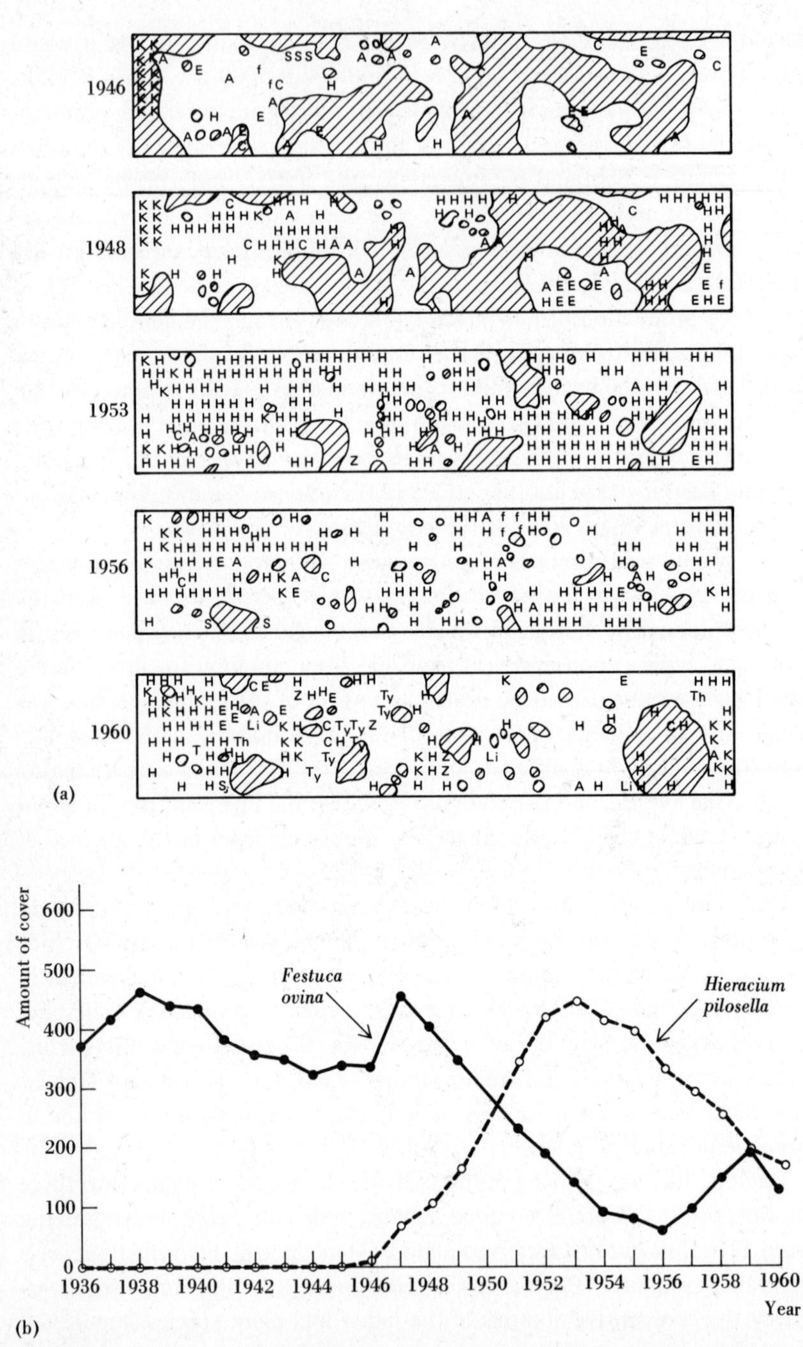

Figure 6.9 Vegetation changes inside a grassland exclosure fenced from rabbits in 1936 in southern England. (a) Vegetation map of part of the exclosure showing how the grass *Festuca ovina* (shaded area) has been replaced by hawkweed *Hieracium pilosella* (symbol H). Other letter symbols stand for minor plant species also affected by grazing. (b) Amount of cover measured from 1936–1960 for these two dominant plant species. (After Watt 1962)

placed by a highly competitive herb *Hieracium pilosella* (hawkweed), which intercepts light and removes water from the soil more efficiently by its root system. The weed *Hieracium* in turn became senescent and was beginning to be replaced by *Festuca ovina* in 1960. If an equilibrium is to be achieved in these grasslands in the absence of rabbit grazing, it may be a dynamic equilibrium with long-term cyclical fluctuations. Plant communities change so slowly that it is difficult to know whether they have stabilized or not if only a few years' data are available.

If biological communities can exist in several stable configurations, humans will typically find some of these configurations more desirable than others. This is clearly true in fish communities and in grazed grasslands and even more noteworthy in communities that contain organisms causing human disease. The biological community in this case is a component of the whole assemblage of plants and animals and is significant to us because it contains humans. One example is the ecology of malaria, a disease caused by microorganisms in the genus *Plasmodium.* Although it is more common and usually more intense in the tropics, malaria can be severe even in the temperate zone (Figure 6.10). Human malaria includes four closely related diseases caused by different species of *Plasmodium.* Humans are infected through the bites of mosquitoes that carry *Plasmodium.* In some instances these parasites can be transmitted to monkeys, but this is rare in nature and malaria is a specifically human disease.

Malaria has probably caused more harm to humans than any other infectious disease, and the control of malaria must rank as one of the great triumphs of applied biology. Malaria was the main cause of infant mortality

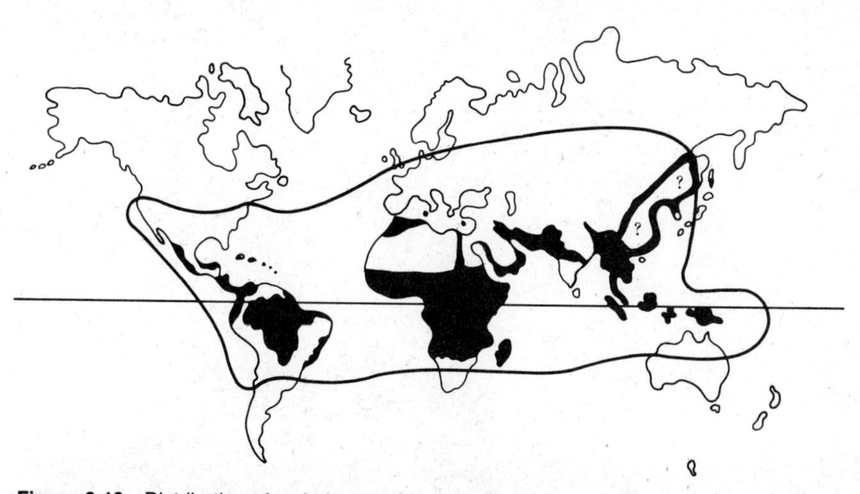

Figure 6.10 Distribution of malaria over the earth. The former limits of malaria are outlined, and transmission still occurs in the black areas. (From Busvine 1975)

in the world until 1940. It played a major role in the decline and fall of the Roman Empire, of Greece, and of early cultures in the tropical and subtropical zones. Until 1940 no one thought that malaria could be eliminated from the earth. Humans, mosquitoes, and malarial parasites had apparently reached a balanced configuration over millions of square kilometers. To disturb this

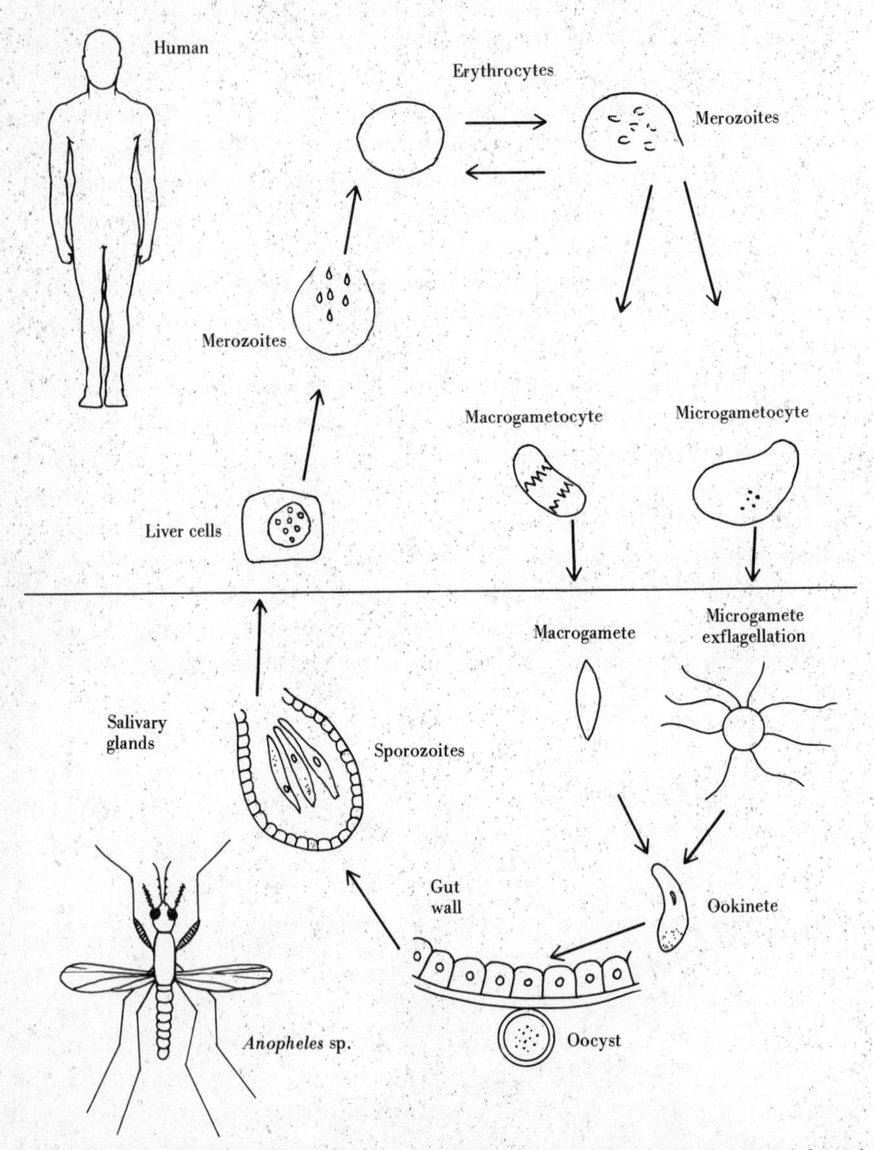

Figure 6.11 Life cycle of the malaria parasite *Plasmodium.* Mosquitoes become infected only by biting people who have malaria, and people become infected only by being bitten by a mosquito carrying *Plasmodium.* (From Aron and May 1982)

balanced configuration seemed impossible, but now we know it can be done and malaria can be reduced greatly or even eliminated.

Malaria is caused by four species of protozoans, smaller than many bacteria, which are inoculated from the mosquito's salivary glands into the human bloodstream (Figure 6.11). The protozoan parasites move from the blood into cells of the liver, where they multiply, and after 12 days burst out of the liver cells back into the blood. In the blood the parasites invade red blood cells (erythrocytes) and multiply, destroying the cells in the process. As the tiny parasites burst from the red blood cells and liberate poisonous toxins from their growth cycle, the human body responds with shivering and fever, the chief symptoms of malaria. These symptoms recur in greater or lesser intensity depending on the strain of malaria and the general health of the infected person. If the person does not die, he or she develops a degree of immunity and as the immunity develops, a new form of the malarial parasite is produced in the blood—the sexual stages of *Plasmodium* (macro- and micro-gametocytes). These sexual stages are infective to any mosquito that bites the person at this time. The male and female *Plasmodium* forms unite in the mosquito's gut and then produce a great number of offspring (sporozoites) that infect the mosquito's salivary glands. The cycle is now complete, and the mosquito can infect the next human it bites. Other mosquitoes in turn become infected with *Plasmodium* only by biting humans who have malaria.

The transmission cycle of malaria is so complex it is a miracle it ever works at all. To control malaria we need to alter this biological community of humans-mosquitoes-parasites to a new configuration in which the parasites become extinct. How can we do this? The simplest suggestion is to kill all the mosquitoes that have any potential contact with people, but Ronald Ross showed in 1911 with a simple mathematical model that this was not necessary.

Consider a single mosquito infected with malaria in a human community. The critical question is whether this one original mosquito will leave on the average *more than one* mosquito offspring that will itself become infectious. If it does, malaria will increase and the system is "above threshold." If it does not, the system is "below threshold," and malaria will decline in the population. What determines the position of these thresholds? Figure 6.12 shows how these thresholds operate. The malaria propagation rate depends on all the complex biological steps shown in Figure 6.11, but we can condense these into two main variables: (1) the proportion of the mosquito population infected with malaria and (2) the proportion of the human population infected. When mosquito populations are dense and individual mosquitoes survive well and bite many people, malaria increases to an upper equilibrium set by human immunity and deaths. But if mosquito densities can be reduced or mosquito survival can be cut down, the system reaches a threshold below which malaria

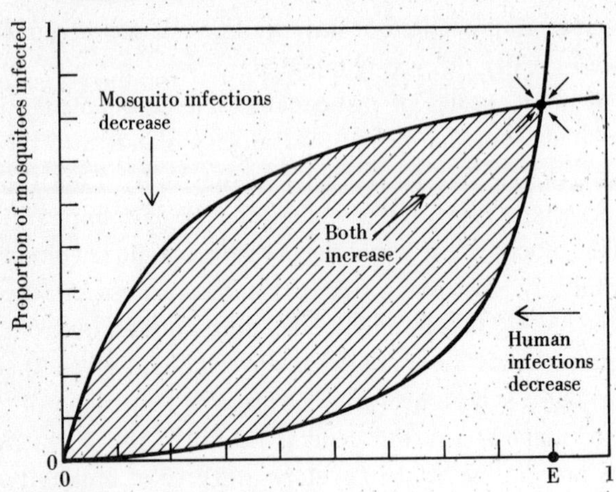

(a) Proportion of humans infected

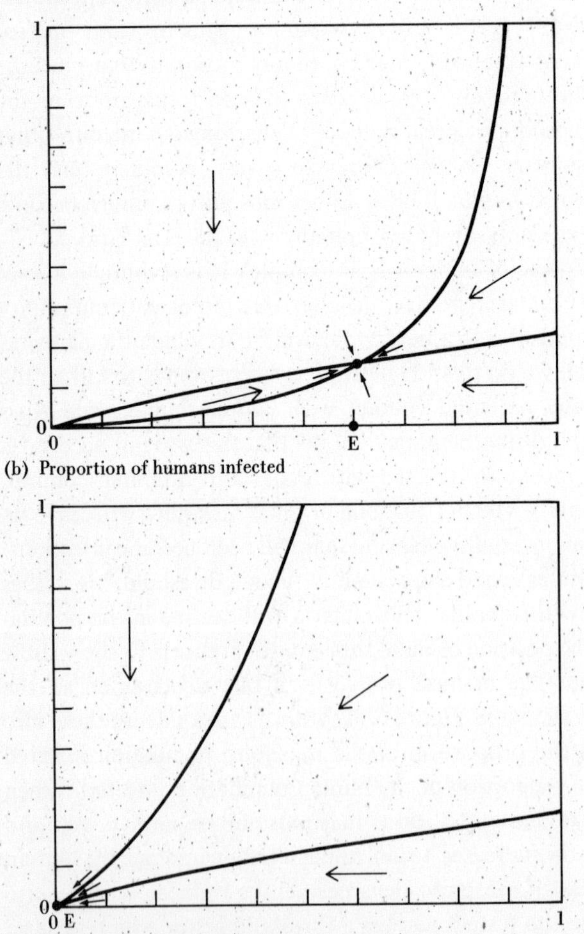

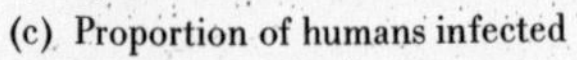

(c) Proportion of humans infected

Figure 6.12 A simple model to illustrate different equilibrium states (E) for the malaria disease system. (a) In areas of endemic, stable malaria there is a high fraction of infected humans and infected mosquitoes. This is the situation in parts of Africa. (b) If public health measures are taken to reduce the number of mosquito bites or to keep mosquitoes from biting infected humans, an equilibrium is formed at a lower infection rate. This is the situation in parts of Asia. (c) If stringent health measures can be taken to reduce mosquito numbers and to prevent contact between humans and mosquitoes, the equilibrium shifts to zero and the disease disappears. This is what has happened in the United States and Europe. (Modified from Aron and May 1982)

cannot propagate and dies out. Malaria can be controlled by reducing mosquito numbers below threshold or by reducing the availability of infected humans. Once the malaria cycle is broken, the system can move to a new configuration with the malaria parasite extinct. In this way malaria, once common in such familiar places as California and the Mississippi Valley, can be eliminated, even though both mosquitoes and people are present in large numbers.

Let us try to draw together a few threads from this discussion of community changes. We started from the simple belief that biological communities that are disturbed—by logging or pollution or fire or whatever—are *resilient,* so that once the disturbance stops, they return to their pristine configuration. But we have found out that this is an ecological myth, that communities will indeed change if disturbed and they may change to a new state that is stable. When we remove the stress, the community may *not* revert to its original condition.

We are still at the infancy stage in trying to manipulate biological communities, and we need to find out what forces can move a community from one configuration to another. At the present time we knock communities about in terribly uncontrolled ways and then, like an infant, sit back in surprise as we watch the results of our disturbances. If we can discover some of the rules by which community changes occur, we would be in a better position to restore damaged communities, like Lake Erie, and to avoid degrading other communities by our disturbances. Biological communities are resilient, but we cannot change them willy-nilly into new configurations and expect that it will be easy and cheap to return them to a more desirable state. The communities we enjoy so much esthetically and economically may be like boulders poised on a mountain slope, easy to dislodge to a lower position but hard to move back up again.

FURTHER READING

Anderson, R. M. (ed.). 1982. *The Population Dynamics of Infectious Diseases: Theory and Applications.* Chapman and Hall, London.

*Desowitz, R. S. 1981. *New Guinea Tapeworms and Jewish Grandmothers: Tales of Parasites and People.* Norton, New York.

*Harrison, G. 1978. *Mosquitoes, Malaria and Man.* Clarke, Irwin, New York.

Laws, R. M. (ed.). 1984. *Antarctic Ecology.* Academic Press, New York.

Likens, G. E. (ed.). 1972. *Nutrients and Eutrophication: The Limiting Nutrient Controversy.* American Society of Limnology and Oceanography, Lawrence, Kan.

National Academy of Sciences (USA). 1970. *Eutrophication: Causes, Consequences, Correctives.* National Academy of Sciences, Washington, D.C.

Pimm, S. L. 1982. *Food Webs.* Chapman and Hall, London.

Zaret, T. M. 1980. *Predation and Freshwater Communities.* Yale University Press, New Haven.

*Highly recommended.

chapter 7

Keystone Species May Be Essential to a Community

Biological communities contain hundreds of species of animals and plants, and we can ask whether all these species are equally important in the community. A species is important in this sense if its removal causes the community to change, either by gaining or by losing some additional species. We thus ask how replaceable a species is, and how much its loss reverberates to other species in a community. This question is critical for conservation because it is asking the biological consequences of extinctions, and the list of rare and endangered species grows yearly.

In any community we can usually identify one or two *dominant* species at each trophic level. By dominant we mean here the species that are most abundant or contain the most biomass. For example, the sugar maple is the dominant plant species in part of the climax forest in eastern North America, and, by its abundance, determines in part the physical conditions of the forest community. Buffalo grass is a dominant perennial in the short-grass prairie of western Kansas. Brown lemmings are the dominant herbivores on the arctic coastal plain of northern Alaska, and wolves are the dominant predators in the boreal forests of central Alaska.

How do species achieve dominance in a given trophic level? There are at least three strategies for achieving dominance. First, *be quick.* A species that can find new habitats quickly and increase in numbers quickly can sometimes be a dominant because it can achieve superior numbers before any competitors arrive. Only a few species can become dominant in this way. Second, *specialize.*

A species that becomes a specialist on a resource that is common and widely distributed can itself become common and an ecological dominant. Third, *generalize.* A species that can use a great variety of foods or other resources can attain numerical superiority, although it will be faced with stiff competition from other species using the same resources. Generalists are dominants only if they have high competitive ability, and most dominant species fit this description.

Since dominant species are so numerous or contain so much biomass, we might expect them to be very important to the community. But what happens when you remove a dominant species from a community? As far as we can tell, often not much happens! The most spectacular case is that of the American chestnut which disappeared from the eastern deciduous forest with no indication of damage to the forest community. Dominant species can be removed from a community because they are involved in strong competition with other species in the same trophic level. When chestnuts died, their place in the forest was taken by oaks, hickories, red maple, and poplars; chestnuts were completely replaceable. Any herbivores that were completely dependent on chestnut would also have disappeared when the chestnuts died, but such specialization seems to be rare, at least in temperate communities.

Thus the dominant species are *not* necessarily essential to a community, contrary to our intuitive feeling that size or abundance should signify "importance." The species that do turn out to be essential are sometimes unexpected ones. Such essential species are called *keystone* species because, like the keystone or central block in an archway, their activities determine the structure of the whole community.

The starfish *Pisaster ochraceous* is a keystone species in the rocky intertidal communities along the west coast of North America. The food web of the rocky intertidal is fairly simple, if we restrict ourselves to the larger invertebrates.

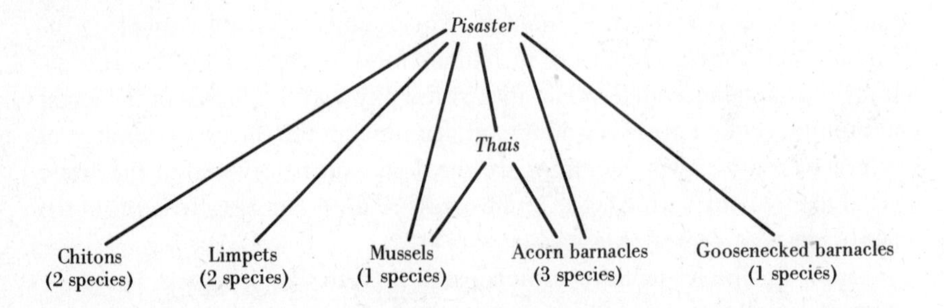

Pisaster is the most common large starfish in the intertidal zone. It weighs 1 to 1.5 kilograms (2 to 3 pounds) and varies from orange to purple in color. It feeds almost exclusively on mussels if given the choice, although it will also eat barnacles, chitons, and snails like *Thais.* The feeding of *Pisaster* on mussels turns out to be critical for all the species in this community.

The key resource in the rocky intertidal community is *space* because organisms cannot survive in this wave-washed environment without a firm place of attachment. Competition for space thus is all important in this habitat. You can see this clearly because when all the rock surface is covered with organisms, they begin to grow on top of one another, as much as their structure will allow. Not all organisms are equal in competition for space, and in the rocky intertidal zone mussels are able to monopolize space. Mussels attach themselves to the rock by a strong byssal thread. When vacant space becomes available, mussels can colonize it in two ways. Larval mussels settle from the plankton during fall and winter. Or larger mussels can migrate as adults by becoming detached (not voluntarily!) and being washed by waves to a new location where they reattach their byssal thread. When other species such as barnacles occupy a space, mussels just grow over them and smother or squeeze them out.

Mussels form a tight band in the intertidal zone (Figure 7.1) and both the upper and the lower limits of this band remain stationary over the years. Larval mussels settle on rocks throughout the whole intertidal zone, but few manage to survive in the lower intertidal. What happens if one removes the major mussel predator, *Pisaster,* from these areas? Robert Paine removed starfish from a rock face on the Washington state coast for six years and

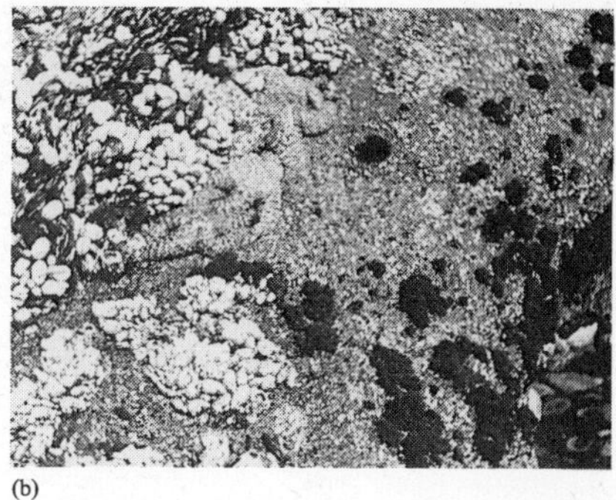

Figure 7.1 The rocky intertidal zone at Mukkaw Bay, Washington. The lower limit of the mussel *Mytilus californianus* has scarcely changed over 20 years in this undisturbed area (a). The starfish *Pisaster ochraceous* (c) feeds on mussels in the lower parts of this zone and thus opens up patches of space for other organisms. Other organisms visible in (b) are tufts of the red alga *Endocladia muricata* and white patches of the goose-necked barnacle *Mitella polymerus.* (Photos courtesy of R. T. Paine)

produced a dramatic effect. Mussels began to extend their range into the lower intertidal. In six years they advanced downward about 1 meter vertically. As they advanced, they took over the rock face and eliminated at least 25 species of large invertebrates and algae (Figure 7.2). The competitively dominant mussels were able to take over all the space in a predator-free zone, and a monoculture of mussels is all that remains. On undisturbed areas starfish can feed only up to a certain level on the shore because they cannot stand long periods of desiccation at low tide. Thus a band forms on the shore at the point of *Pisaster* penetration.

Starfish cannot eliminate mussels from the intertidal zone because mussels have a *refuge* high in the intertidal where starfish cannot feed. But there is yet another way in which mussels can elude starfish predation—by growing too large for starfish to handle. Starfish kill mussels by pulling their two shells apart, but if a mussel grows to a large size, a starfish cannot exert enough force to open it. The trick, of course, is to survive long enough to reach this large size, and by chance a few individuals make it. Since large mussels produce large numbers of eggs, these few individuals can contribute much to the reproductive rate of a mussel population.

The starfish *Pisaster* is thus a *keystone* species in the middle zone of the rocky intertidal because when it is removed, the entire community changes dramatically in composition. Keystone species can thus be recognized only by removal experiments, and they are to be looked for in communities with two features: (1) Some primary producer or consumer is capable of monopolizing a basic resource like space and excluding other species. (2) This resource monopolist is itself preferentially consumed or destroyed by the keystone predator or herbivore.

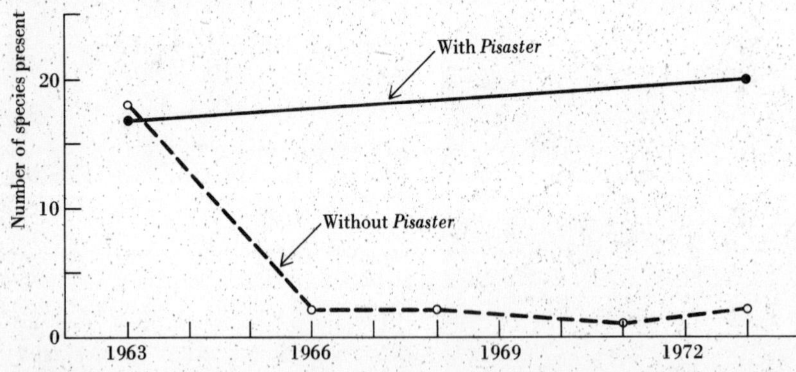

Figure 7.2 Changes in the number of species of the larger invertebrates and algae on a horizontal rock surface at Mukkaw Bay, Washington, after the experimental removal of the predatory starfish *Pisaster*. The removal experiment began in July 1963. In the absence of starfish predation the mussel *Mytilus californianus* is able to take over all the space and eliminate the other species. (Data from Paine 1974)

These starfish removals have been repeated in New Zealand and in Chile where different species of starfish eat other species of mussels. The results were identical—in the absence of starfish predators mussels of various species tend to monopolize the rocks of the upper intertidal zone. The keystone predator effect thus seems to be general to many rocky coasts of the temperate zone throughout the world.

How common are keystone species in natural communities? No one knows, yet it is clearly important that we find out from a conservation point of view. Keystone species may be more common in aquatic communities than in terrestrial ones. Let us look at another aquatic example.

Off the east and west coasts of North America kelp *(Laminaria)* beds form an important subtidal community. Kelp beds can be destroyed by sea urchins that graze on the kelp and other algae that comprise these underwater forests. The sea urchin *Strongylocentrotus droebachiensis* is nearly spherical in shape and has well-developed jaws, capable of eating large seaweeds at a great rate. It seems to live in three different types of subtidal communities. In dense kelp beds urchins may live for many years at low density (1 per 10 square meters). They live in crevices from which they never seem to emerge. Their food is largely algal detritus.

When urchins become more abundant (30 to 100 per square meter), they form dense aggregations in the open and feed on large kelp plants. The aggregations may be massive and they may march as a front through kelp beds, attacking the bases of the plants and eating everything as they go.

After the dense aggregations have completely destroyed the kelp bed, the urchins do not die. They persist on the bare areas of rocky substrate variously called barren grounds or coralline flats. Little grows on these areas except the flat coralline algae, and sea urchins feed on drift algae and diatom films that cover the rocks. Urchins grow slowly and rarely reproduce on these barren grounds, but they survive for many years.

Rocky bottoms in the subtidal zone can exist as rich kelp forests or as bare barren grounds depending on the abundance of sea urchins. What controls the abundance of sea urchins? Is there a keystone predator in this marine system? The answer is yes, but the keystone predator is not the same in all communities in the different oceans. Compare the situation off eastern Canada with that in California and Alaska.

The key predator on sea urchins off eastern Canada and New England is the lobster *(Homarus americanus).* Lobsters attack sea urchins by turning them over and breaking their calcareous skeleton near the base. Lobsters have been subjected to heavy fishing because of their commercial value, and the reduction of lobster densities by fishing may be responsible for allowing sea urchins to explode in numbers and destroy the kelp beds (Figure 7.3). Since

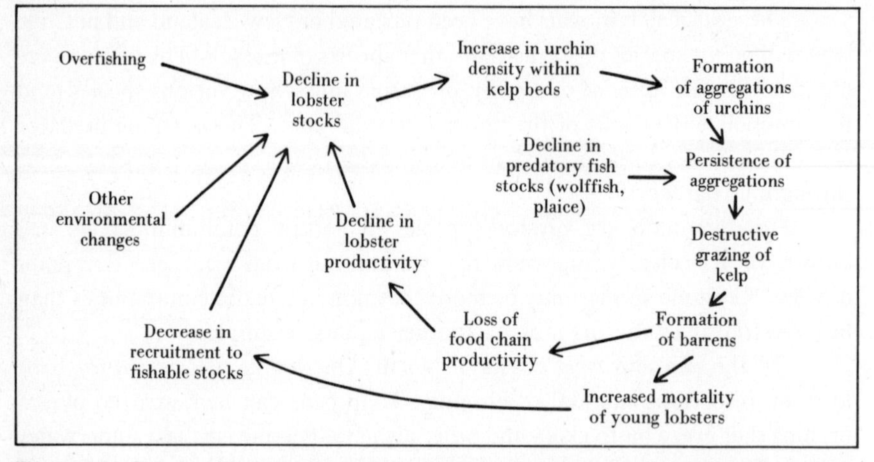

Figure 7.3 Relationship between declining lobster populations and the destruction of kelp beds by sea urchin grazing. (After Wharton and Mann 1981)

the kelp beds are the major producers in the subtidal zone and provide shelter and food for a great variety of fish and other invertebrates, the loss of the kelp forests causes a great change in the community. The lobster is so far the only commercially exploited species off the Atlantic coast that seems to act as a keystone species, but the unfortunate truth is that we know so little about the communities of most exploited species that we could not recognize keystone effects if they did occur. In this case the productivity of lobsters depends directly on the productivity of the kelp beds, and the management strategy for the lobster fishery off New England should be directed at the restoration of kelp beds.

The key predator on sea urchins in the North Pacific is the sea otter *(Enhydra lutris)*. Sea otters were once abundant around the North Pacific rim from Japan to California. They feed primarily on sea urchins when they are available. In the absence of such predation urchin populations overgraze their algal food sources and prevent kelp establishment. Experimental urchin removal in subtidal quadrats in southeastern Alaska produced rapid colonization by kelp, and the sites became dominated by *Laminaria groenlandica.* The reintroduction of sea otters has resulted in a dramatic increase in kelp biomass, and this has in turn increased the abundance of near-shore fishes (Figure 7.4). Once they reduce the abundance of sea urchins, sea otters must spend more time feeding because the fish that are their alternative prey are more difficult to capture than sea urchins. The removal of sea otters by fur trappers in the nineteenth century thus had a significant impact on the organization of subtidal communities. Sea otters are now protected and as they spread back along the Pacific coast, a reversal is occurring and dense kelp beds are appearing again.

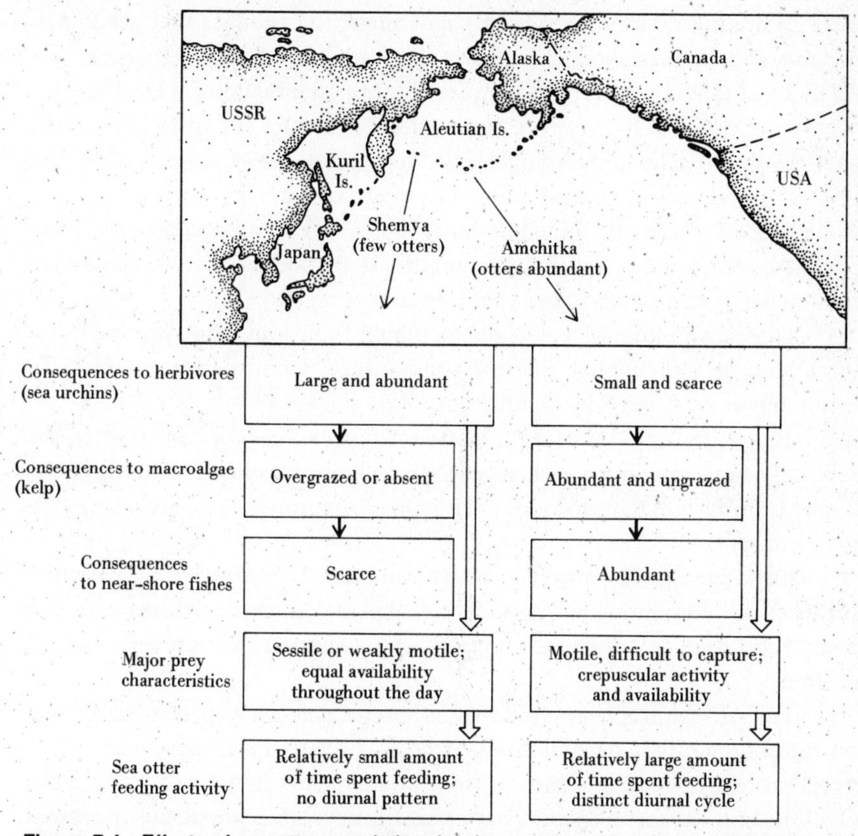

Figure 7.4 Effects of sea otter population density on sea urchins, kelp, and fishes in the Aleutian Islands of Alaska. (Modified from Estes et al. 1982)

Other mechanisms may intervene to reduce sea urchin numbers once they are high. Epidemic disease, associated with warm ocean temperatures, broke out among the sea urchins along the Nova Scotia coast in 1980 and spread in 1981 and 1982. This epidemic virtually eliminated sea urchins to a depth of 25 meters along at least 500 kilometers of coastline, and algae have regenerated within 3 years so that extensive kelp beds again line the shores as they did 20 years ago. When and if sea urchins recover from this epidemic, we may see the whole cycle repeated again.

The importance of keystone species in natural communities is part of an overall acrimonious verbal battle between two opposing schools of community ecologists. One school, led by Michael Gilpin, Jared Diamond, and Jonathan Roughgarden of California, argues that most communities are controlled by competition between species and are in a state of equilibrium, saturated with as many species as they can hold. The other school, led by Dan Simberloff of Florida, John Wiens of Colorado, and Joe Connell of California, argues that most communities are controlled by environmental changes that are unpredictable and by predation and disease. Communities, Simberloff argues, are rarely in a state of equilibrium because of all the fluctuations in weather that occur in nature and because predators and diseases are not spread uniformly over the world. The keystone species concept is an important component of this second view that natural communities are continually fluctuating. Keystone predators prevent the community from suffering the consequences of complete competition. Natural communities that contain keystone species are thus richer systems than they would otherwise be.

These two alternative schools of thought are important in a practical sense because they predict different consequences from species removals. In communities controlled by competition, the removal of one or a few species will rarely be noticed. But in communities controlled by keystone species, the removal of one species (like the starfish) can have dramatic consequences. It is clearly important to us to know what kind of communities we are manipulating in nature.

Keystone species may not be as common in terrestrial communities because they may rarely have one species that is competitively dominant and able to monopolize critical resources. Large mammals may provide the best examples.

The African elephant is a keystone species that has a spectacular impact on the community. The African elephant is a relatively unspecialized herbivore, but relies on a diet of browse from woody plants supplemented by grass. Over the last 30 years elephants have transformed dense woodlands in eastern and central Africa into more open vegetation types (Figure 7.5), and this has been widespread enough to be called "the elephant problem." By their feeding activities, elephants destroy shrubs and small trees and push woodland habitats toward open grassland or open savannah. Large trees can be uprooted or broken off or killed by elephants feeding on the bark. The net result is that elephants are killing trees faster than the trees regenerate. As more grasses invade the woodlands, the frequency of fires increases, which accelerates the conversion of woods to grassland. This conversion works to the elephant's disadvantage, however, because grass is not a sufficient diet for elephants, and they begin to starve as the woody species are eliminated. Other ungulates, such as zebra and wildebeest, eat the grasses and are favored by the elephant's activities.

There are two quite different views about the underlying causes of the elephant problem, and we need to look at them in some detail because they

Figure 7.5 Elephant damage to woodland vegetation in Africa. (a) Undisturbed *Terminalia* woodland in North Bunyoro National Park, Uganda, outside the geographical range of elephants. (b) Open grassland that was formerly *Terminalia* woodland in the south part of Kabalega Falls National Park, Uganda. Elephants have destroyed the woodland. (c) Part of a herd of 340 elephants, Kabalega Falls National Park, Uganda. Dead trees were killed by elephants. (Photos courtesy of R. M. Laws)

(a)
(b)
(c)

produce contradictory recommendations for management of elephants. The first view is the *natural equilibrium hypothesis.* It states that elephants and trees exist in a stable equilibrium that is occasionally upset (Figure 7.6). The upset can be produced by any number of ecological forces including human activity, and the most important recent event may have been the compression of elephants into the national parks as the land around the parks was converted to agriculture (Figure 7.7). Other upsets could be the elimination of hunting, fires, epidemic diseases, or habitat changes caused by climatic change. The net result is that elephants are allowed to increase for a time, but their overgrazing results in starvation and an eventual slow return to the stable equilibrium.

The second view is the *limit cycle hypothesis.* It states that there is no attainable natural equilibrium between elephants and forests. Instead the system cycles back and forth, the elephants increasing in density as they thin out the forest and declining in density as the trees become too sparse (see Figure 7.6). When elephants decline to low numbers, the trees are able to regenerate, and the cycle begins once again. This cycle is also open to human interference. For example, humans can stop the cycle in the low phase by partially clearing

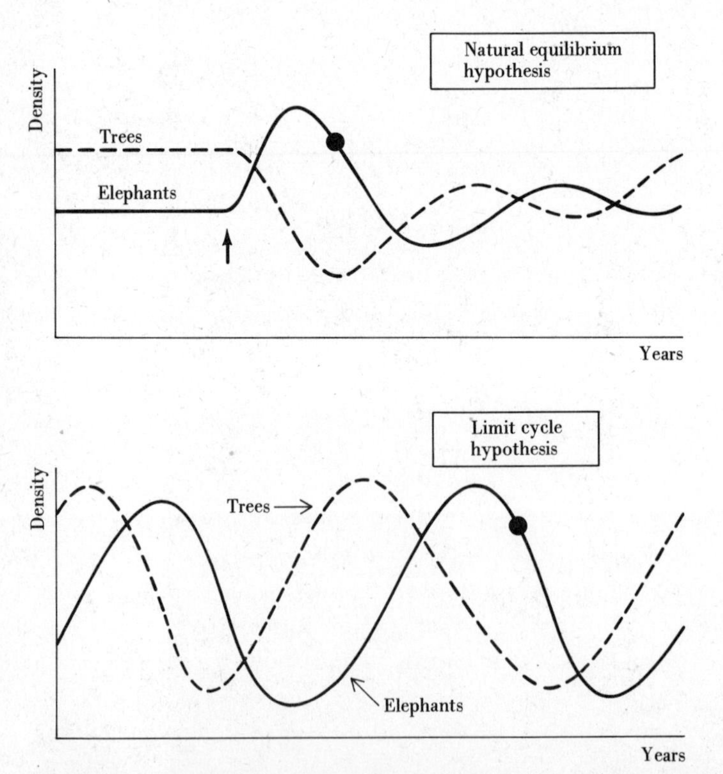

Figure 7.6 Two alternative views of the elephant problem in Africa. The arrow in the upper graph indicates an ecological disturbance, and the dot indicates the current position of the African populations.

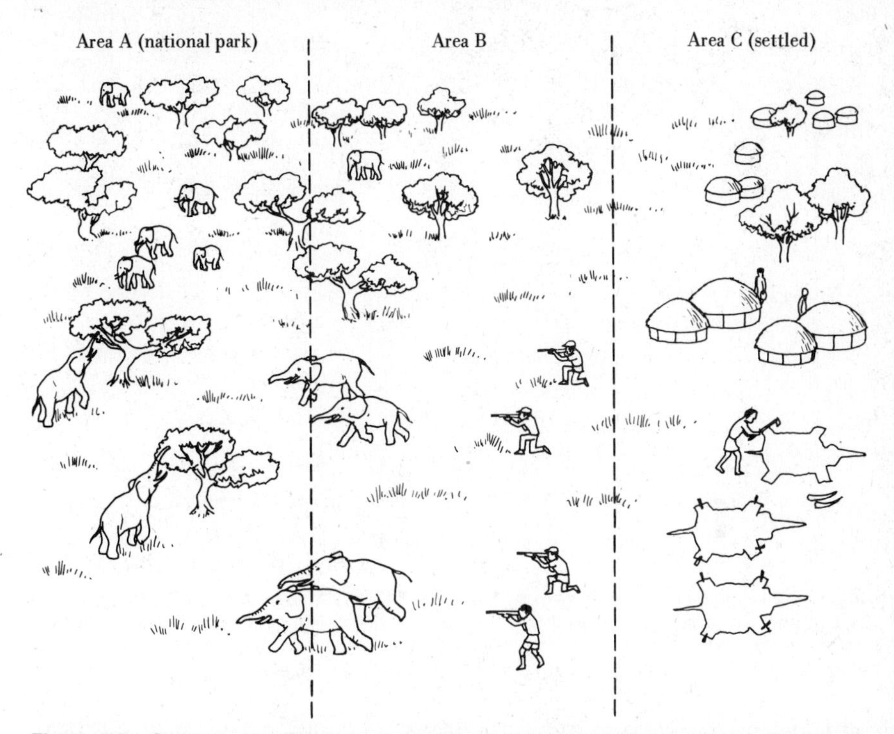

Figure 7.7 Compression. This is one possible reason why elephants often suffer from a population problem in national parks. When area C is settled, most of the elephants are killed. People then begin to infiltrate the band of country (area B) bordering the national park (area A), killing some elephants and harassing others, which then flee into the safety of the park, which already has its full complement of elephants. (After Eltringham 1982)

land and starting fires, which act to prevent tree regeneration. The important point to note is that these two hypotheses are exact reverses of each other. The cycle hypothesis implies that the elephant problem is characteristic of the system, but that the system could be modified by humans to achieve an artificial equilibrium. The equilibrium hypothesis begins from the assumption that the system is characterized by a natural equilibrium that can be disturbed by human activity to produce the elephant problem.

Some information on the time scale of the elephant problem can be obtained from the age structure of trees on which the elephants feed. Unfortunately, only a few of the main food trees lay down annual growth rings so that we can age them easily. Graeme Caughley sampled baobab trees in the Luangwa Valley of Zambia and obtained the data shown in Figure 7.8. Elephants find young baobabs irresistible, and so eliminate most of the young trees. The recruitment of young baobab trees has declined progressively over the last 140 years, and the inference is that elephants have been affecting the forest seriously for at least 100 years. The elephant problem is not new! Since large baobab trees can live to be 1000 years old, Caughley's data are consistent

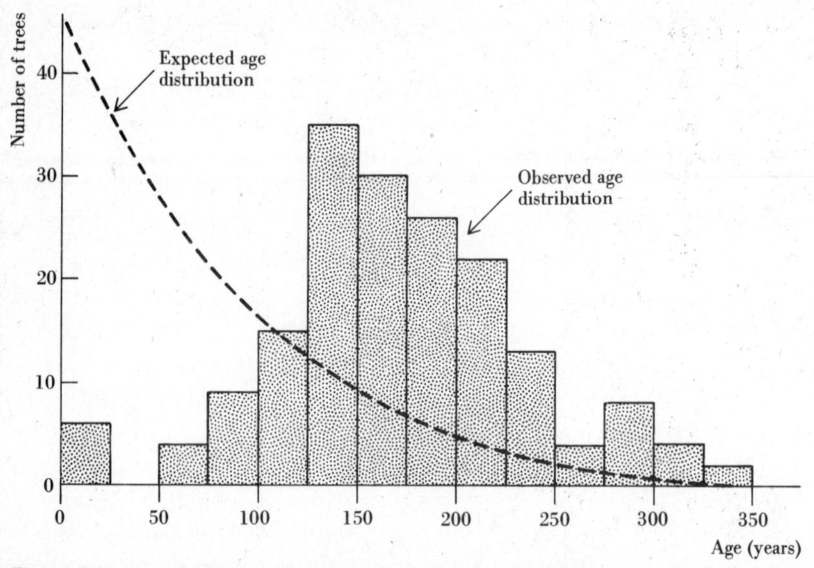

Figure 7.8 Age distribution of a random sample of 173 baobab trees in the Luangwa Valley of Zambia. The expected curve shows the general shape the age distribution should take if there were no elephant damage to young baobab trees. Only a few young trees have been able to survive during the last 140 years. (After Caughley 1976)

with the limit cycle hypothesis, with the minimum of the elephant cycle possibly having occurred 140 years ago. But we can also interpret these tree data as being consistent with the natural equilibrium hypothesis, if we assume some ecological disturbance about 140 years ago. The tree data do not allow us to decide which hypothesis is correct, but they do help define the time scale of the problem.

If the time scale of the elephant problem is of the order of 150 to 200 years, it will take a long time to distinguish the population trends predicted by the two hypotheses. The consequences to elephant management are great. If the natural equilibrium hypothesis is correct, the selective culling of "problem" elephant populations will bring the system back to equilibrium much more rapidly, and the need for management would disappear. If the limit cycle hypothesis is correct, the selective reduction of elephant populations will only speed one along the cycle, and the problem will never go away. Elephants could be held at a low density, but only by constant management efforts.

This dichotomy of approach was shown well in the Tsavo National Park of Kenya. Within 20 years of its establishment after World War II, elephants in Tsavo had destroyed large parts of the dense woodland communities and converted these into grassland savannah. During the late 1960s two polarized views developed. Some scientists recommended artificial reduction of the elephant population, while others, including the Kenya National Parks' administration, advocated noninterference on the assumption that the system would eventually correct its imbalance.

An unusually severe drought in 1970 and 1971 led to high elephant mortality. In the eastern part of Tsavo National Park about 6,000 elephants died out of a total population of 25,000. This severe mortality fell primarily on the young animals (up to 5 years) and on the females of reproductive age. The net result has been reduced pressure on the vegetation and reduced reproductive capacity of the elephant population. The drought, of course, also affected the vegetation and by itself caused tree mortality. Grasses are much more resistant to drought, and drought also increases the chances of severe fires. The net result of the drought on the whole community is not clear.

The effect of elephants in converting woodland to grassland is analogous to the effect of starfish in converting mussel beds into an open, complex community. The physical size of elephants makes their effects on the forest so conspicuous that we fail to notice or to study the other effects that follow from the changed habitat. One of the paradoxes of ecology is that the largest and most dramatic ecological phenomena are simultaneously the easiest to observe and the hardest to study. We cannot manipulate elephants in the same easy way we can manipulate starfish, and the time scale of the experiments we need to do on the elephant problem exceeds our human life span. The issues involved in the elephant problem are sufficiently important to the long-term conservation of the wildlife of Africa that we must set up the needed experiments, even if they take a hundred years to complete. In politics taking the long view means looking four years ahead, but in wildlife conservation we must realize that *the larger the animal, the larger the problem and the longer the time frame we must adopt.* We should not shrink from setting up observation and data collecting experiments that take a hundred years to complete, and it will be a credit to our ecological wisdom if we can report to our children: We have begun.

We began this chapter asking how we could determine which of the many species in a community were essential to the existence of the community. We have seen that at least in some communities there are keystone or critical species whose presence or absence reverberates throughout the community. These keystone species need not be common in an undisturbed community. But ecologists do not know how many keystone species there are in natural communities. Are they rare and unusual or common but mostly undiscovered? It is possible that many or even most of the plant and animal species we see in nature could be eliminated with little measurable effect, but it is not wise to eliminate species whose roles in nature are incompletely understood. Ecologists cannot at present predict the consequences of any extinction and in this, as in all of our interactions with nature, we should be true conservatives.

FURTHER READING

Diamond, J., and T. J. Case (eds.). 1986. *Community Ecology.* Harper & Row, New York.

*Hanks, J. 1979. *A Struggle for Survival: The Elephant Problem.* Country Life Books, London.

Price, P. W., C. N. Slobodchikoff, and W. S. Gaud. 1984. *A New Ecology: Novel Approaches to Interactive Systems.* Wiley, New York. (Especially Chapters 13 and 17.)

Steele, J. H. 1974. *The Structure of Marine Ecosystems.* Harvard University Press, Cambridge, Mass.

Strong, D. R., Jr., D. Simberloff, L. G. Abele, and A. B. Thistle. 1984. *Ecological Communities: Conceptual Issues and the Evidence.* Princeton University Press, Princeton, N.J.

*Highly recommended.

chapter 8

Natural Systems Recycle Essential Materials

Recycling was not invented by environmental activists in the 1960s but was evolved by biological communities millions of years ago. Through the long history of evolution, plant and animal communities have learned the hard lesson that every schoolchild learns after opening his or her first bank account: Input must equal output in the long run. Of course, input may exceed output, and our bank account can grow, but the converse can never occur over long time periods. Recycling then is sensible indeed, and we need to look at some of the details of how natural communities recycle essential materials and some of the difficulties they face.

Plants and animals require two basic physical provisions: *energy* and *materials.* Energy in ecological systems originates from the sun, and all of the energy needed must be captured by green plants through photosynthesis. The sun's energy is captured in plant carbohydrates, proteins, and lipids and then used by animals through the food chain. Energy is not recycled by organisms, but is ultimately used for activity or growth and lost as heat. The continued input of solar energy keeps the ecological machine running, and organisms can afford to use energy because there is always more on the way.

Materials are completely different. They are not provided anew each day and must be conserved. Conservation can be achieved only by recycling, and so we need to find out how chemical materials cycle in nature. Two different types of cycles occur (Figure 8.1). Gaseous elements—carbon, nitrogen, oxygen, and hydrogen—all circulate in *global cycles* because they are exchanged

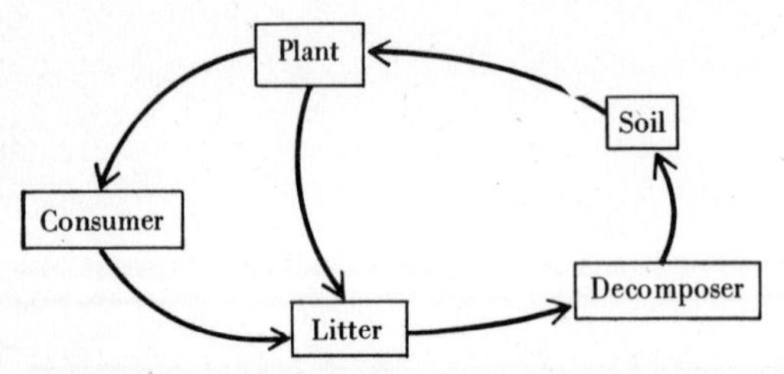

(a) Local cycles of P, K, Ca, Mg, Cu, Zn, B, Cl, Mo, Mn, and Fe

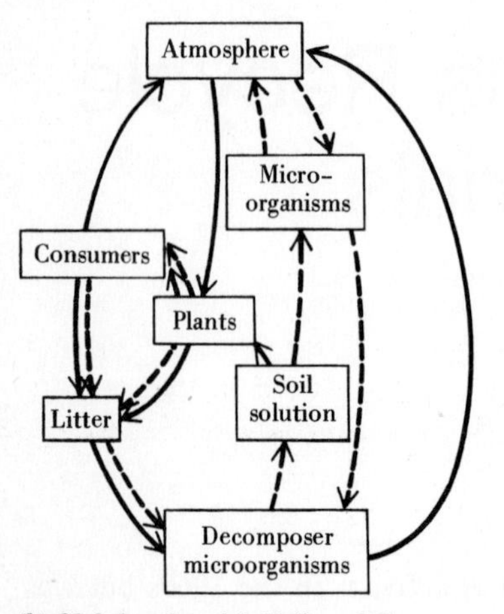

(b) Global cycles of C, N, O, and H

Figure 8.1 Generalized diagrams of two types of nutrient cycles: local cycles and global cycles. The dashed lines in the global cycles refer to nitrogen only. (After Etherington 1975)

between the air and biological organisms. Long-distance transfers are common, and the oxygen you are breathing today is part of a pool or reservoir having many distant sources. Solid elements—phosphorus, potassium, calcium, magnesium, copper, zinc, boron, chlorine, molybdenum, manganese, and iron—all circulate in *local cycles* and have no mechanism for long-distance transfer.

To analyze any particular nutrient cycle, we need to measure first the amount of the nutrient in each "box" of Figure 8.1 and second the flow rate of the nutrient between boxes. Once we have completed this description, we can try to determine the main factors affecting the transfer of nutrients in the community. Nutrient cycles have been studied in many biological systems, particularly in systems that humans exploit.

The nutrient dynamics of a forest are summarized in Figure 8.2. Each box in this figure can be studied in detail and broken down further. One fact is clear, even from this simplified diagram: *Biological communities are not closed systems.* Animals move from one community to another, drainage water transports dissolved materials to adjacent lakes and streams, and logs removed

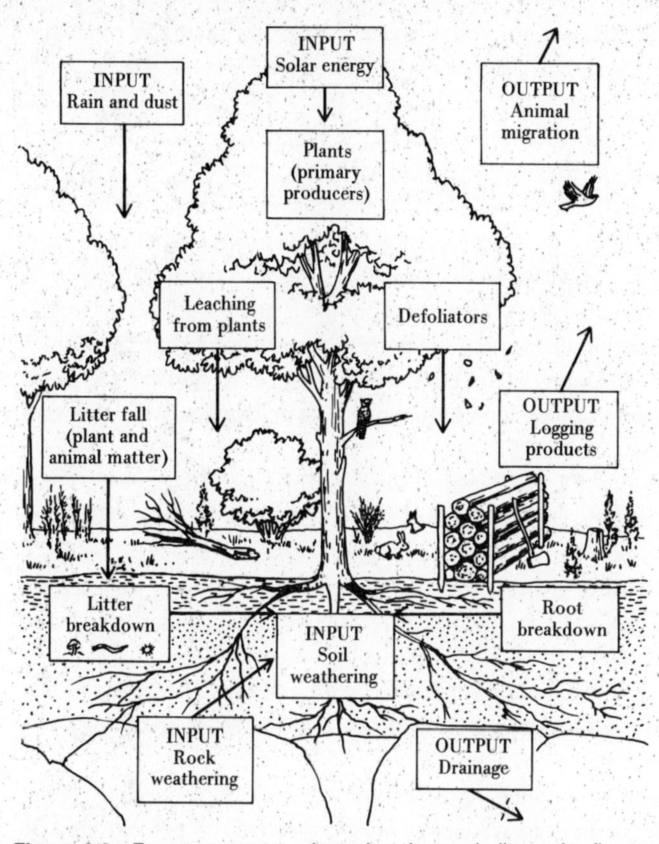

Figure 8.2 Forest ecosystem dynamics. Arrows indicate the flows of matter and energy. (After Ovington 1962)

from a forest contain a significant investment in nutrients. These losses must be counterbalanced by nutrient gains from rain, dust, and litter and from soil and rock weathering if there is to be no long-term decline in forest productivity. These gains and losses have been studied in detail in only a few cases.

One of the most extensive studies of nutrient cycling in forests has been carried out at the Hubbard Brook Experimental Forest in New Hampshire. The Hubbard Brook forest is a nearly mature, second growth hardwood ecosystem. The area is underlain by rocks that are relatively impermeable to water, and hence all runoff occurs in small streams. The area is subdivided into several small watersheds that are distinct yet support similar forest communities, and these watersheds are good experimental units for study and manipulation.

Nutrients enter the Hubbard Brook forest ecosystem in precipitation. The precipitation input was measured in rain gauges scattered over the study area. Nutrients leave the ecosystem primarily in stream runoff, and this loss was estimated by measuring stream flows. For most dissolved nutrients the streamwater leaving the system contains more nutrients than the rainwater entering the system. About 60 percent of the water that enters as precipitation leaves as stream flow; most of the remaining 40 percent is transpired by plants or evaporated. The chemical composition of the precipitation and the stream discharges changes little from year to year.

Annual nutrient budgets can be calculated for watersheds in the Hubbard Brook system, based on the difference between precipitation input and stream outflow. Eight chemicals show net losses from the ecosystem: calcium, magnesium, potassium, sodium, aluminum, sulfate, silica, and bicarbonate. Three chemicals showed an average net gain: nitrate, ammonium, and chloride. If we assume that these nutrient budgets should be in equilibrium in this undisturbed ecosystem, the net losses must be made up by chemical decomposition of the bedrock and soil.

After obtaining background information for the intact watersheds, Gene Likens at Cornell and F. H. Bormann at Yale studied the effects of logging on the nutrient budget of a small watershed at Hubbard Brook. One 15.6-hectare watershed was logged in 1966, and the logs and branches were left on the ground so that nothing was removed from the area. Great care was taken to prevent disturbance of the soil surface to minimize erosion. For the first three years after logging the area was treated with a herbicide to prevent any regrowth of vegetation. This deforested watershed was then compared with an adjacent intact watershed.

Runoff in the small streams increased immediately after the logging, and annual runoff in the deforested watershed was 41 percent, 28 percent, and 26 percent above the control in the three years after treatment. Detritus and debris in the stream outflow increased greatly after deforestation, particularly two to three years after logging. Correlated with this was a large increase in streamwater concentrations of all major ions in the deforested watershed. Nitrate concentrations in particular increased 40 to 60 times over the control values (Figure 8.3). For two years the nitrate concentration in the streamwater of the deforested site exceeded the maximum safe level recommended for drinking water. Average streamwater concentrations increased 417 percent for calcium, 408 percent for magnesium, 1558 percent for potassium, and 177 percent for sodium in the two years after deforestation.

From 1969 onward the deforested area was allowed to recover, and the speed of recovery was measured over the next seven years (Figure 8.4). Some of the chemical and stream flow measures returned to normal within three to four years of recovery, but the logged area had not made a complete recovery after 7 years. Stream flow was still higher on the logged area, and consequently more calcium and potassium were still being lost from the logged site after 7 years than from the forested site.

These results were obtained from an experimentally deforested area where wood was not removed and where the forest floor was essentially

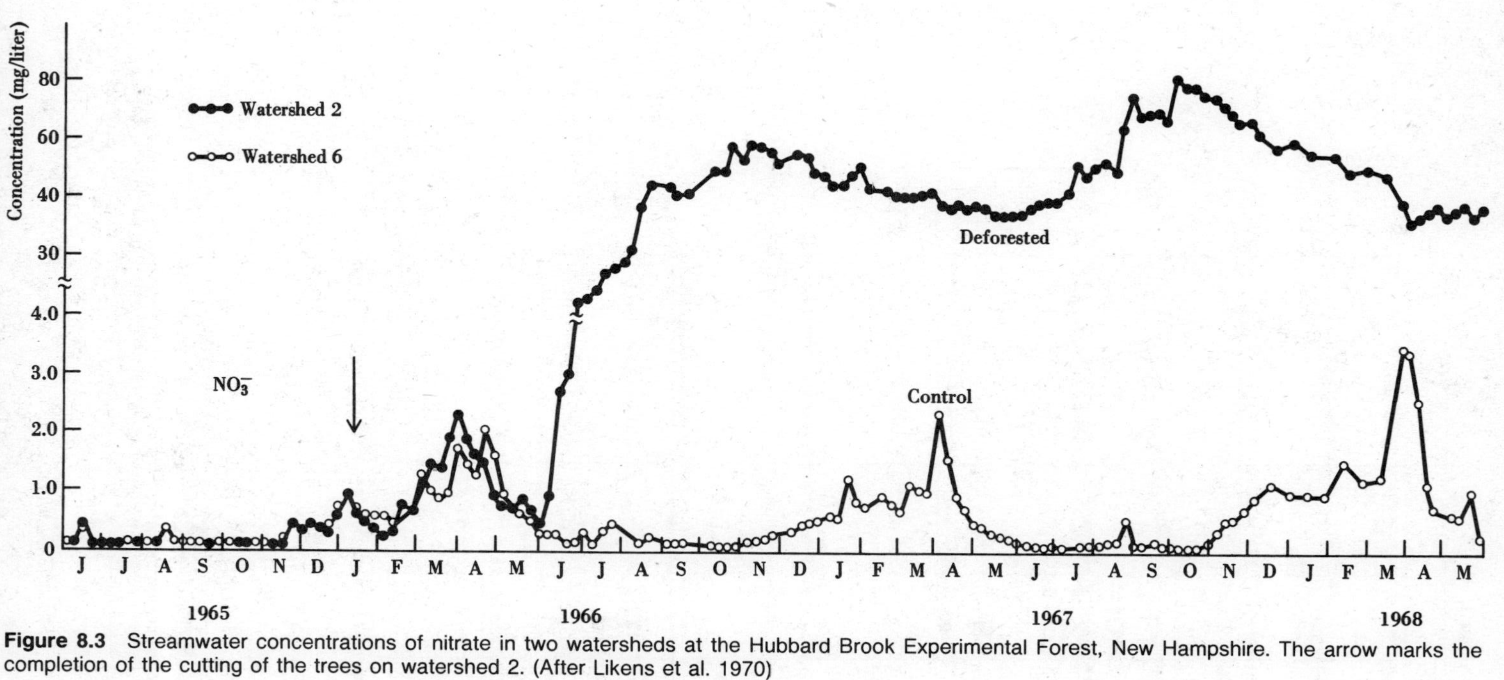

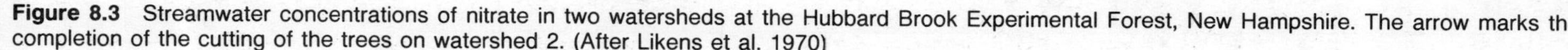

Figure 8.3 Streamwater concentrations of nitrate in two watersheds at the Hubbard Brook Experimental Forest, New Hampshire. The arrow marks the completion of the cutting of the trees on watershed 2. (After Likens et al. 1970)

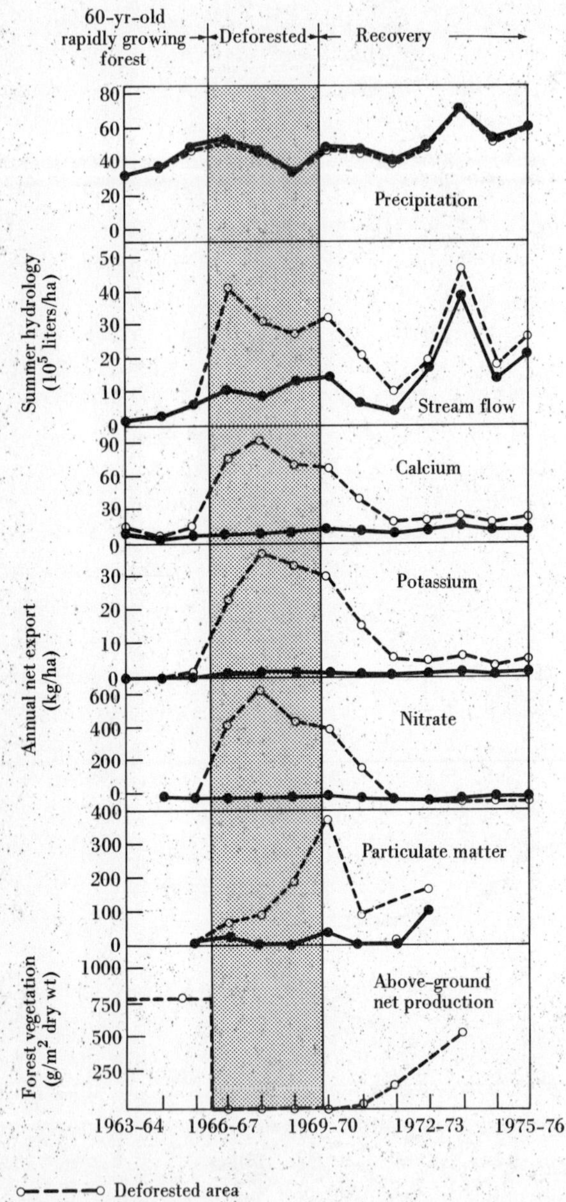

Figure 8.4 Effects of deforestation on water flow and nutrient losses in a northern hardwood forest at the Hubbard Brook Experimental Forest, New Hampshire. A 60-year-old, rapidly growing forest was cut during the autumn of 1965, maintained bare of vegetation for three growing seasons, and then allowed to revegetate during the growing season of 1969. Extensive nutrient losses followed deforestation. (After Likens et al. 1978)

undisturbed. To see what the results of commercial clear-cutting might be, Likens and his associates (1978) studied eight clear-cut sites in the White Mountains of New Hampshire. The same general results were found. Nutrient losses in stream runoff were reduced by the immediate vegetative regrowth that occurred after the logs were removed, but additional nutrients were lost in the logs that went to market, so that the total nutrient losses from the commercially logged sites exceeded those from Hubbard Brook.

Nutrient losses after clear-cutting can be recovered if the site is not disturbed by severe soil erosion. Rock weathering restores many of the important elements like calcium and potassium. Nitrogen is captured from the air by certain bacteria and algae and converted to nitrate in the soil. At least 60 to 80 years are required for nutrient recovery at Hubbard Brook, and consequently the present forestry practice of 110- to 120-year rotation for cutting should allow the forest to recover completely between cuttings.

By looking at commercial forestry from a viewpoint of nutrient cycling we can help to recommend sound management procedures in forestry. For example, bark is relatively rich in nutrients and hence lumbering operations ought to be designed to strip the bark from the trees at the field site and not at some distant processing plant. Similarly, the debris of branches and leaves left on the logged area, although unsightly, represents part of the nutrient capital of the site that will be recycled as it decomposes. Ecologically sound harvesting practices are not always esthetically pleasing or the most convenient methods.

Most of the nutrients lost from clear-cut areas end up in streams and lakes where they become part of the aquatic pollution problem we discussed earlier (Chapter 6). Undesirable losses from some biological communities can become undesirable gains for another community. These flows of nutrients show graphically how interdependent communities can be.

Natural communities seem to exist in a state of long-term nutrient balance, but some exceptions stand out. Peatlands are areas in which plant production exceeds decomposition so that organic matter accumulates and with it a deposit of nutrients. In peatlands input exceeds output. Why do peatlands form, and how do nutrients cycle in peatlands?

Peatlands cover about 3 percent of the earth's land surface. Peat is produced in any area where moisture is in excess supply throughout the growing season. Excessive moisture selects for certain kinds of plants because it retards decomposition. As leaves and stems die and fall to the soil surface, they are only partly decomposed, and litter begins to build up. The critical boundary in a peat bog is the depth of the water table in summer. Above this boundary, the *active* zone, plant materials are broken down more rapidly because oxygen is present. Below this boundary, the *inactive* zone, plant materials are broken down slowly because no oxygen is present (microorganisms use it up more quickly than it can diffuse in). Of all the plant matter that falls to the peat bog surface, about 80 to 95 percent is lost by decomposition before it moves into the inactive zone.

Peatlands can be produced in any climatic zone, but they are most extensively distributed in cold, wet regions (Figure 8.5). Canada and the Soviet Union contain about 75 percent of the world's peat. Decomposition is stopped by freezing winter temperatures, and drainage is impeded on many areas underlain with permafrost (permanently frozen subsoil). However, peatlands can be extensive outside of arctic regions. Coal deposits laid down in the Carboniferous period 400 million years ago are just extensive peat bogs developed in a warm climate with high rainfall.

The plants growing in peatlands vary considerably depending on the geographical area. The typical peatland of north temperate climates contains at least one species of *Sphagnum* moss which often acts as a keystone species in these communities. The genus *Sphagnum* contains the most important peat-forming group of plants in the world. *Sphagnum* mosses have no roots but grow upward so that the upper stem elongates while the lower stem gradually dies and enters the peat. *Sphagnum* grows slowly, and a peat bog dominated by *Sphagnum* will have a low annual production of new plant material, much lower in fact than arctic tundra or desert areas. *Sphagnum* has two other peculiarities that strongly affect the peatland: It absorbs cations

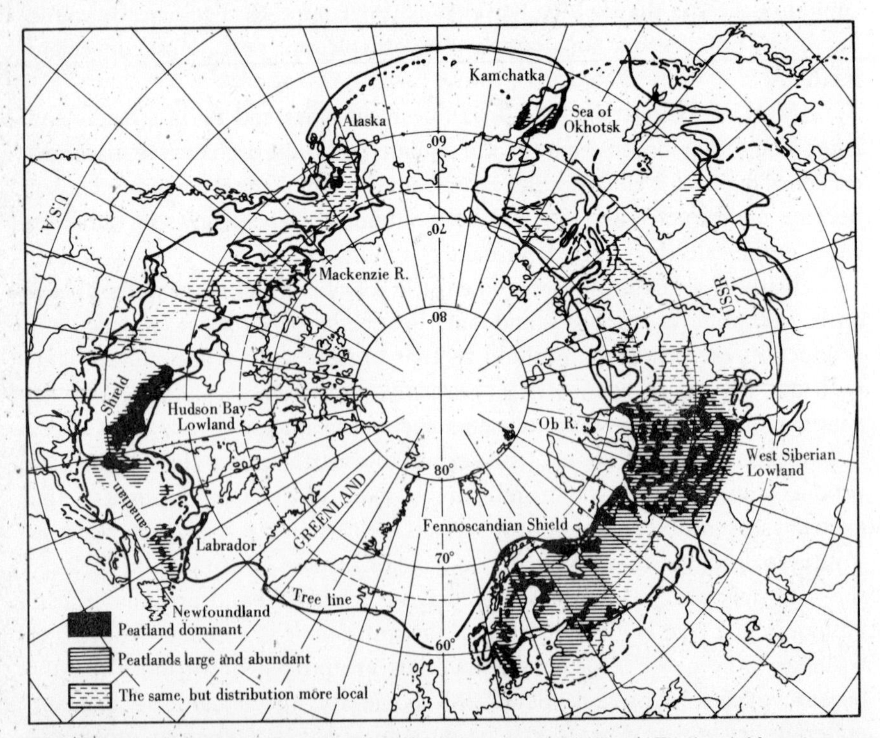

Figure 8.5 Areas of abundant peatland in the Northern Hemisphere. The boreal forest zone, just to the south of the tree line, contains most of the world's peatlands. (From Sjors 1961)

(Ca^{+2}, K^{+}, Mg^{+2}) from the soil by exchanging hydrogen ions (H^{+}) so that the peat becomes very *acidic;* and it contains little plant nutrients, so that peatlands become *nutrient-poor.*

The waterlogged, inactive region of a peat bog typically becomes anaerobic as the bacteria and fungi in the peat use up all the oxygen. Little decomposition can occur in the absence of oxygen, and thus the deeper layers of peat are held as a nearly unchanging storehouse of partly decomposed organic matter. The upper region of a peat bog is a zone of decomposition by fungi and bacteria because oxygen is available from the air, at least when the water table falls for short periods. Decomposition is most rapid right at the surface of the bog. Some plant chemicals are decomposed quickly, but the resistant chemicals that make up the plant cell walls (lignins and cellulose) are broken down slowly. For this reason one can find almost complete "fossil" leaves and branches buried deep in a peat bog.

Peat accumulation depends on the balance of the production of new plant material and the decomposition that occurs throughout the active and inactive zones. It thus varies with the climatic conditions of the area and the length of the growing season, but in most cases peat accumulates very slowly

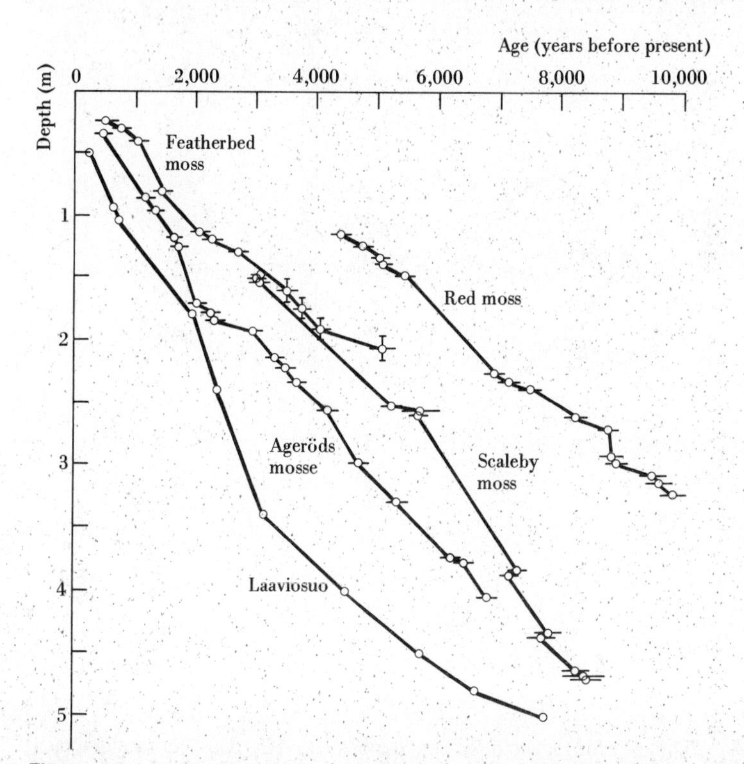

Figure 8.6 Age-depth profiles for five peat bogs from Europe. Age is measured as years before present by the carbon 14 method. Featherbed moss, Red moss, and Scaleby moss are in north and central England, Ageröds mosse in south Sweden, and Laaviosuo in south Finland. (After Clymo 1984)

indeed. An average figure for north temperate peatlands would be 20 to 80 centimeters of peat per 1000 years (Figure 8.6). As the climate changes, so does peat production. Peat was accumulating about three times faster 2000 years ago than it is today. Peat is an important fuel in Scotland, Ireland, and the Soviet Union. It is a *renewable* resource, but its rate of renewal is slow and the rate of harvesting needs to be controlled.

Most peat bogs in the northern temperate zone seem to reach an equilibrium at a depth of 5 to 10 meters in which the rate of addition of plant matter at the surface is balanced by decomposition at all depths. This equilibrium depth is dependent on drainage, vegetation, and temperature and some peat deposits can grow very deep for time periods of more than 50,000 years (Figure 8.7).

Nutrient cycling in peatlands is particularly interesting because of the

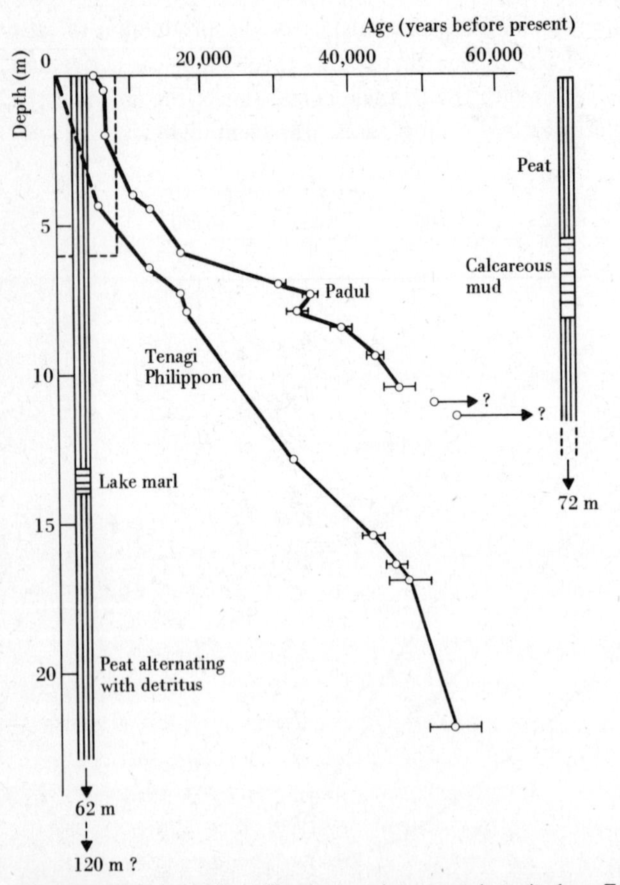

Figure 8.7 Age-depth profiles for two deep peat deposits from Europe. The stratigraphy at the right and the upper curve are for a deposit at Padul in the Sierra Nevada of southern Spain. The deposit is 72 meters deep; the top 10 meters are shown. The stratigraphy at the left and the lower curve are for a deposit at Tenagi Philippon, Greece. The deposit is at least 62 and perhaps 120 meters deep. (After Clymo 1984)

reduced decomposition. The simplest case to study is the peat bog in a depression with no surface drainage. As the peat accumulates, the bog is cut off from the mineral soil below, and so nutrients cannot enter from rock weathering. All nutrients thus come in precipitation or from the air. Precipitation contains nutrients like calcium or magnesium because these elements are left behind in tiny dust particles when sea spray evaporates. Precipitation also picks up other dust particles from the air so that one can measure significant amounts of all the necessary elements in rainfall.

Nitrogen and phosphorus are the two elements in shortest supply in many peat bogs. There is an enormous capital of nitrogen in peat but most of it is completely locked up in complex organic compounds that plants cannot use for growth. Plants in peat bogs survive either by having a low nitrogen requirement and slow growth (as *Sphagnum*) or by utilizing nitrogen directly from the air. Some bog plants have root nodules containing bacteria that convert nitrogen from the air into nitrate, which the plant then uses for growth. A few specialized plants get their nitrogen from the air in a different way—by trapping insects in their leaves. These carnivorous plants may have sticky or rolled leaves; they have enzymes to break down trapped insects so that they can recover nutrients by reversing the normal food chain.

Phosphorus is also scarce in many peat bogs, and plants survive only by having low phosphorus requirements. There is no source for phosphorus except rainwater or groundwater.

When the water table falls because the climate changes or because humans drain the bog, oxygen becomes available throughout the peat profile and decomposition accelerates. Given time and oxygen, peat will break down into humus, and consequently peat is widely used as organic manure to improve the structure of agricultural soils. After drainage, the peatland itself may be used as an agricultural soil, but the resulting soil will be deficient in nitrogen and phosphorus and may also be deficient in other nutrients depending on the chemical composition of the peat. The bogs of western Ireland, for example, cannot be converted to pasture or to forest unless a phosphate fertilizer is used.

Complete nutrient budgets have been constructed for only a few peatlands. The peat-covered landscape of western Ireland is one good example. Blanket bogs cover 3 percent of the total area of Ireland, but these peatlands are a recent development. Until about 4000 years ago western Ireland was covered by forests of Scot's pine and oak. The present bogs are near sea level but low in nutrients. All of the nutrient input comes as rain or dust from the air, and all of the output occurs in drainage waters, so it is relatively easy to get a nutrient budget for the bog. The accompanying table gives these data for a Glenamoy blanket bog in western Ireland for a three-year period from 1969 to 1972, with an estimate of nutrients stored irrecoverably in deep peat each year. It shows that the bog system is not in a steady state where input equals output, but is running down. The only nutrients in excess are magnesium and potassium which are supplied in sea spray. Since peat is actively accumulating in western Ireland, a large supply of nutrients is being stored as peat each year. But if the supply of nutrients for plants is running down, as the data suggest, this must result in a decrease in plant growth and peat production, so that sooner or later erosion will begin to take away the nutrient capital stored in the peat.

Nutrient	Nutrient gain or loss (mg/m²)	Nutrients stored in peat (mg/m²)
Calcium	−2061	6364
Magnesium	+2343	9091
Potassium	+225	4546
Ammonium nitrogen	−1692	3200 (total nitrogen)
Nitrate nitrogen	−752	
Phosphorus	−20	2045
Iron	−371	Unknown

Source: Data from Moore, Powding, and Healy, 1975, p. 336.

Because nutrients are so low already, forest cannot get reestablished on blanket bogs in Ireland without human help. By fertilizing and draining the bog, foresters can establish trees. The whole plant community changes once fertilizer is added and nutrient cycling accelerates. Plant species characteristic of the bog are replaced by species with higher nutrient requirements; two-thirds of the bog plants disappear. Insects become more common in the forest sites. Trees utilize more soil water than bog plants and this causes a further drop in the water table, accelerating the decomposition of the underlying peat. Whether this reversal of peatland formation toward coniferous woodland will continue without continued intervention by humans adding nutrients and draining the soil is an open question that only future research can answer.

Peat bogs are an example of infertile soils because of the shortage of nitrogen and phosphorus. Soil fertility varies greatly in different parts of the globe and this profoundly affects crop growth. Large areas of the Northern Hemisphere have been glaciated and the soils, derived from till in which the bedrock has been pulverized, are very fertile with a high availability of nutrients. Areas of volcanic activity can also have rich soils. But in much of the world soils are old, highly weathered, and basically infertile. For example, the continents derived from Gondwanaland—Australia, South America, and India—have large areas covered with old, poor soils. The vegetation supported on these soils has developed efficient nutrient use by recycling within the plant and by leaf fall and reabsorption.

Australian soils are typical of old, highly weathered soils and contain almost no phosphorus. Eucalyptus trees are adapted to grow on soils of low phosphorus content. If you measure the size of trees and the nutrient content of forests growing on good soils from temperate areas and from Australian sites on poor soils, you will find no suggestion that eucalyptus trees growing on poor soils are smaller or have lower amounts of nutrients than forests in other countries, with the single exception of phosphorus. Eucalyptus have only one-half to one-fifth the amount of phosphorus in their tissues as do Northern Hemisphere trees.

Plants growing in nutrient-poor soils ought to use nutrients more efficiently than plants in fertile soils. Plants from nutrient-poor habitats achieve this efficiency by being better at nutrient absorption than plants from nutrient-rich habitats. One consequence of this efficiency is that forest productivity may be high on soils with low nutrient levels.

Forest ecosystems employ two types of nutrient cycling strategy. The *oligotrophic* strategy occurs on nutrient-poor soils, like those of the Amazon Basin, and the *eutrophic* strategy occurs on nutrient-rich soils. In oligotrophic systems most of the nutrients are in the humus layer of the soil, and this layer of fine roots and humus is critical for nutrient cycling and nutrient conservation. The soil itself contains few nutrients. By contrast, in eutrophic systems the soil as well as the humus layer is rich in nutrients. In the temperate zone, where most forest research has been done, forests are usually of the eutrophic type. Tropical forests are more often oligotrophic.

Productivity and nutrient cycling do not differ significantly in oligotrophic and eutrophic forests, *as long as these ecosystems are not disturbed.* Much of the nutrient capital in tropical oligotrophic forests is in the trees. But when the forest is cleared for agriculture and the trees are harvested, the nutrient-poor oligotrophic systems quickly lose their productive potential, while the nutrient-rich eutrophic ones do not. Once the humus and root layer on top of the mineral soil is disturbed in oligotrophic systems, the mechanism of efficient nutrient recycling is lost and the nutrients are leached out of the system by rainfall. Thus tropical rain forest land, the classical oligotrophic system, cannot be used for crop production unless all nutrients are supplied in fertilizers.

Nutrient cycling can be disrupted by human activities so that nutrients are lost and ecological systems downgraded. But nutrient loss is not the only possible consequence of human activities. Some of our worst impacts are from nutrient additions. We have already discussed the consequences of eutrophication in lakes (Chapter 6). Acid rain is another example.

Human activity through the combustion of fossil fuels has altered the sulfur cycle more than any of the other nutrient cycles. While human-produced emissions of carbon dioxide and nitrogen are only about 5 to 10 percent of the level of natural emissions, we produce about 160 percent of the level of natural sulfur emissions. One clear manifestation of this alteration of the sulfur cycle is the widespread problem of acid rain in Europe and North America. Acid precipitation is defined as rain or snow that has a pH of less than 5.6.

Dissolved carbon dioxide from the air will normally produce rainwater that is slightly acid, but pH values below 5.6 are caused by strong acids (sulfuric acid, nitric acid) that originate as combustion products from fossil fuels. Over large areas of western Europe and eastern North America, annual pH values of precipitation average between 4.0 and 4.5, and individual storms may produce acid rain of pH 2 to 3 (Figure 8.8).

Ore smelters and electrical generating plants have increased emissions during the past 50 years (Figure 8.9). Smelters and generating plants have built taller stacks to reduce local pollution at ground level, but the taller stacks have exported the pollution problem downwind.

The effects of acid rain on the environment are the subject of much current research. Some effects are clear already. Freshwater ecosystems seem to be particularly sensitive. In areas underlain by granite rocks, which are highly resistant to weathering, acid rain is not neutralized in the soil, so lakes and streams become acidified. Lakes in these bedrock areas typically contain soft water of low buffering capacity. Thus bedrock can be an initial guide to sensitive areas. The Precambrian Fennoscandian Shield in Scandinavia, the Canadian Shield, all of New England, the Rocky Mountains, and other areas are thus potential trouble spots.

The clearest effects of acid precipitation have been on fish populations in Scandinavia and eastern Canada. Fish populations have been reduced or

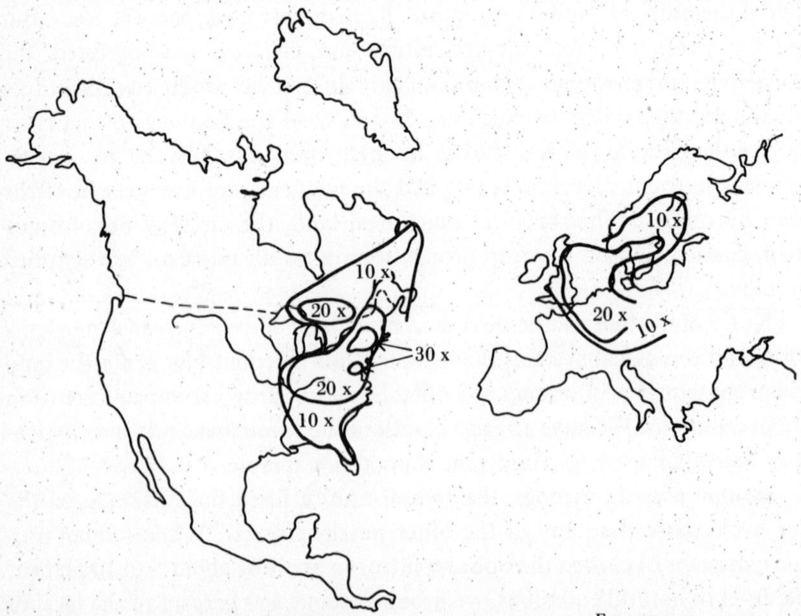

Figure 8.8 Distribution of acid precipitation in North America and Europe. Areas designated 10 x, 20 x, and 30 x receive 10, 20, and 30 times more acid in precipitation than expected if the pH were 5.6. (After Likens et al. 1981)

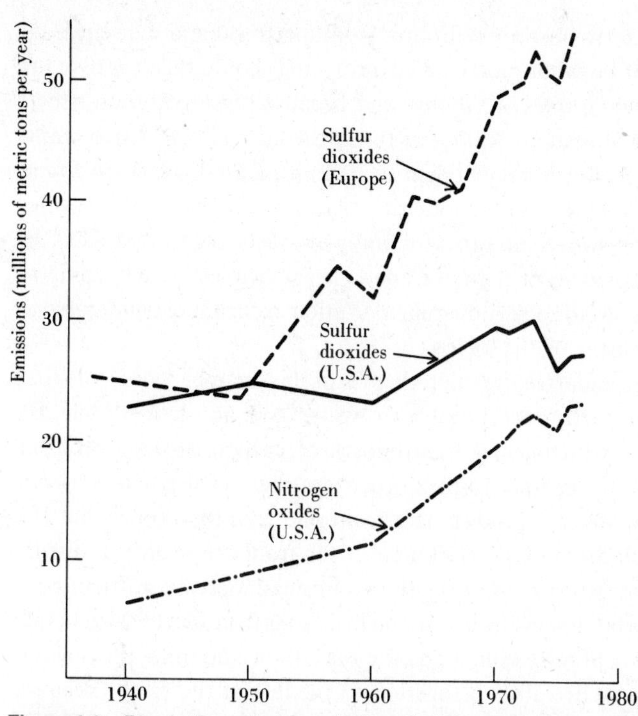

Figure 8.9 Rise in the emissions of sulfur dioxides and nitrogen oxides in the United States and Europe, 1937–1977. The curves reflect increases in the burning of fossil fuels and the smelting of ores. Sulfur and nitrogen may be converted into strong acids in precipitation. (After Likens et al. 1979)

eliminated in many thousands of lakes in southern Norway and Sweden once these waters have fallen below pH 5.

The effect of acidic precipitation on terrestrial ecosystems is more complex and difficult to unravel. Highly acidic precipitation can reduce production in terrestrial plants by direct toxicity or leaching of nutrients from the foliage. The rate of decomposition in the soil may be slowed by lower pH, and this could interfere with nutrient cycling of other chemical elements. In some cases productivity of terrestrial systems might *increase* if a limiting nutrient like nitrogen or phosphorus is added in the precipitation. Acid rain could also improve soils that are low in sulfur. The main point is that acidic precipitation has the potential to change nutrient cycling in natural ecosystems in a great variety of ways we cannot yet understand, much less predict. We should not continue this aerial bombardment of ecosystems in the naive belief that nutrient cycles have infinite resilience to human inputs.

Widespread damage to forest trees in central Europe has been observed since the early 1980s. This damage is particularly severe in West Germany and Switzerland and seems to be a joint product of acid rain damage and ozone pollution. Ozone is produced by the action of sunlight on nitrous oxides and hydrocarbons that are emitted by automobile exhausts. There is as yet no clear understanding of exactly how air pollution is killing trees, and consequently there is no agreement on what measures governments should take to solve the problem. But meanwhile trees are dying, and because these pollution effects require several years to actually kill trees, forest damage will get much worse before it gets better again. Surveys taken in Britain in 1985 and 1986 found 20 percent of all Scot's pines with moderate or severe damage to the needles in parts of the country where no problem had previously been recognized. In Quebec and other provinces of eastern Canada sugar maples are beginning to die in large numbers. Acid rain and ozone pollution are leading candidates to be the ecological disaster of the 1980s.

Let me try to summarize the main themes of this chapter. Nutrients that are needed for life on earth pass through air, water, soil, and living tissue. By following molecules of nitrogen or phosphorus or carbon through all their travels, we can begin to see the complex process of these cycles. Humans can disrupt the cycles by diverting substances from one area to another, but the consequences of such disruptions often appear far from the source of disturbance. Logging a watershed high in the mountains can increase nutrient concentrations in waters far downstream. Sulfate emissions from electrical generating plants can cause acid rain, killing aquatic life a thousand miles downwind. The message is that we are all tied together to the fate of the earth. Because of the world scale of nutrient cycles, we are linked in time to past and future biological communities and in space to communities all over the world. There are no more islands.

FURTHER READING

*Elsworth, S. 1984. *Acid Rain in the U.K. and Europe.* Pluto Press, London.

Gore, A. J. P. 1983. *Mires: Swamp, Bog, Fen and Moor.* A. *General Studies.* Elsevier, Amsterdam.

Jordan, C. F. 1985. *Nutrient Cycling in Tropical Forest Systems.* Wiley, New York.

Likens, G. E. (ed.). 1981. *Some Perspectives of the Major Biogeochemical Cycles.* Wiley, New York.

Moore, P. D., and D. J. Bellamy. 1973. *Peatlands.* Elek Science, London.

Pomeroy, L. R. 1973. *Cycles of Essential Elements.* Dowden, Hutchinson and Ross, Stroudsburg, Pa.

Ulrich, B., and J. Pankrath (eds.). 1983. *Effects of Accumulation of Air Pollutants in Forest Ecosystems.* Reidel, Dordrecht.

*Highly recommended.

chapter 9

Climates Change, Communities Change

We grow used to the daily and seasonal changes in weather, and we view these changes as variations within a constant climate. Our faith in the constancy of climate may be shaken by observing that there are wet years and cold years, but we assume that over the long run the apparent deviations will all average out. This assumption of climatic constancy is completely wrong. One of the major themes to emerge from the last 30 years of scientific research is that climate changes on all time scales.

Evidence of changing climates comes from a variety of sources. Let us start with recent history in which we have a written record of change. Figure 9.1 shows a thousand years of temperature changes from Iceland. Thermometer data go back only 140 years in Iceland and before this one must rely on indirect evidence—reports of drift ice floating by the island. Drift ice is a thermometer of the North Atlantic. The earliest reports are from A.D. 825 when Irish monks visited Iceland and found no ice there even in winter. To reach pack ice they had to sail one day's journey to the north. There was no record of drift ice off Iceland for at least 400 years until the 1200s when severe winter pack ice moved south and cut off the old Norse colonies in Greenland, causing them to be abandoned. From about 1500 to 1900 the best years in Iceland were only as good as the worst years the Vikings had experienced during the tenth and eleventh centuries. From about 1890 to 1950 a great

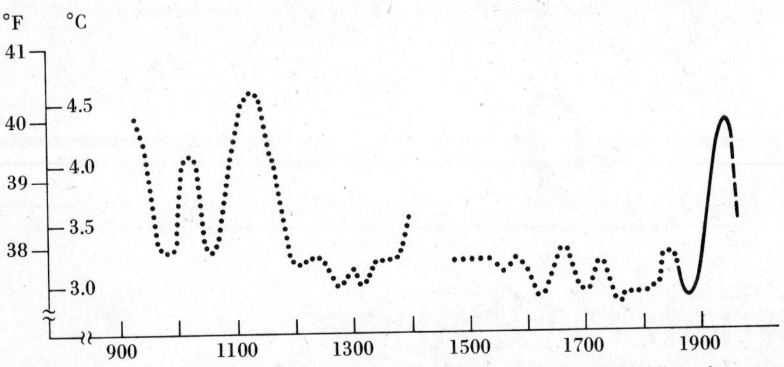

Figure 9.1 A thousand years of Icelandic temperatures. The dotted line shows estimated average temperatures (with the early part of the fifteenth century omitted because of insufficient data). The solid line shows temperatures based on thermometer readings. The dashed line shows the change in recent years for the Northern Hemisphere. (From Bryson and Murray 1977)

warming occurred, followed by a cooling during the last 25 years. A message clearly emerges from this Icelandic saga: *The climate of the present century is not typical of the previous thousand years.*

To extend our vision about climatic change back further, we can use pollen grains trapped in lake sediments. Pollen grains in sediments are fossils of older plant communities. Because different species of plants have pollen that differ in shape and size (Figure 9.2), ecologists can reconstruct plant communities back in time. In combination with radiocarbon dating of sediments, we can determine the timing of community changes over the globe.

Fossil pollen studies have proved successful at describing the sequence of plant communities over the past 30,000 years. Because of the Pleistocene ice sheets in the Northern Hemisphere, climatic and vegetational shifts occurred around the globe. Figure 9.3 on page 132 shows the pollen record for Rockyhock Bay in North Carolina, an area of the eastern United States about 500 kilometers south of the line of maximal glacial advance. In this bay about 5 meters of sediment has been deposited over the past 30,000 years. From 30,000 to 21,000 years before the present (B.P.) temperate forests of oaks *(Quercus)*, birches *(Betula)*, and pines *(Pinus)* occupied the area. From 21,000 to 10,000 B.P. boreal forest was present, with spruce *(Picea)* and northern pines, and the climate was colder and drier than at present. Deciduous forests, with a predominance of oak, replaced the boreal forests about 10,000 B.P. Swamp forests began to develop about 7,000 B.P., featuring blackgum *(Nyassa)*, cedar *(Cuppressaceae)*, magnolia *(Magnolia)*, and red maple *(Acer rubrum)*. The swamp forests were essentially modern by 4,000 years ago.

Most of the plant and animal communities of the polar and temperate zones are still recovering from the last episode of Pleistocene glaciation that

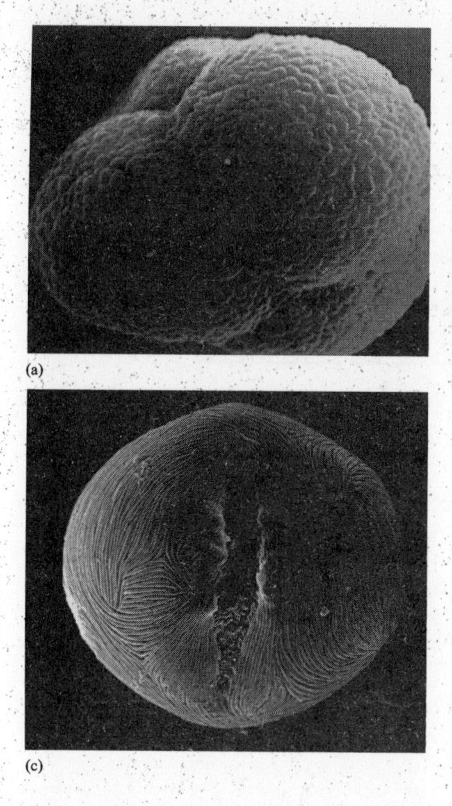

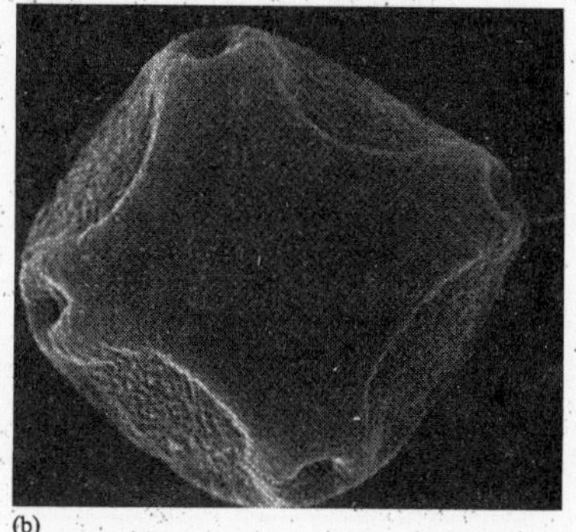

Figure 9.2 Scanning electron microscope photos of three types of pollen grains. (a) Heather, *Calluna vulgaris* (.03 × .04 mm). (b) Alder, *Alnus glutinosa* (.03 mm). (c) Mountain ash, *Sorbus ancuparia* (.04 mm). (Photos courtesy of P. D. Moore)

ended about 15,000 years ago. We can look for evidence of climatic fluctuations most easily at the tree line. The present polar limit of trees in northern Canada and Alaska has changed dramatically since the Ice Age ended and is still changing today. Figure 9.4 on page 133 shows the changes in pollen rain for a small lake that is now surrounded by arctic tundra 70 kilometers north of the present tree line in the Northwest Territories of Canada. Since ice began receding about 14,000 years ago, this area has been occupied by tundra plants, then by boreal forest, and finally by tundra again. These changes have occurred because of slow but systematic changes in the amount of solar energy reaching the surface of the earth. In northern Canada there has been a general shift in the polar limits of trees, which were 200 to 300 kilometers farther north 1000 years ago than they are today.

Another technique employed to trace recent climatic fluctuations is the analysis of tree growth. Trees in the temperate and polar zones are particularly affected by how good the growing season is. Figure 9.5 on page 134 illustrates how temperature variation from year to year affects the growth of bristlecone pine *(Pinus longaeva)* needles. Trees in the temperate and polar regions lay down annual layers of wood (seen as rings in a cross section). The better the

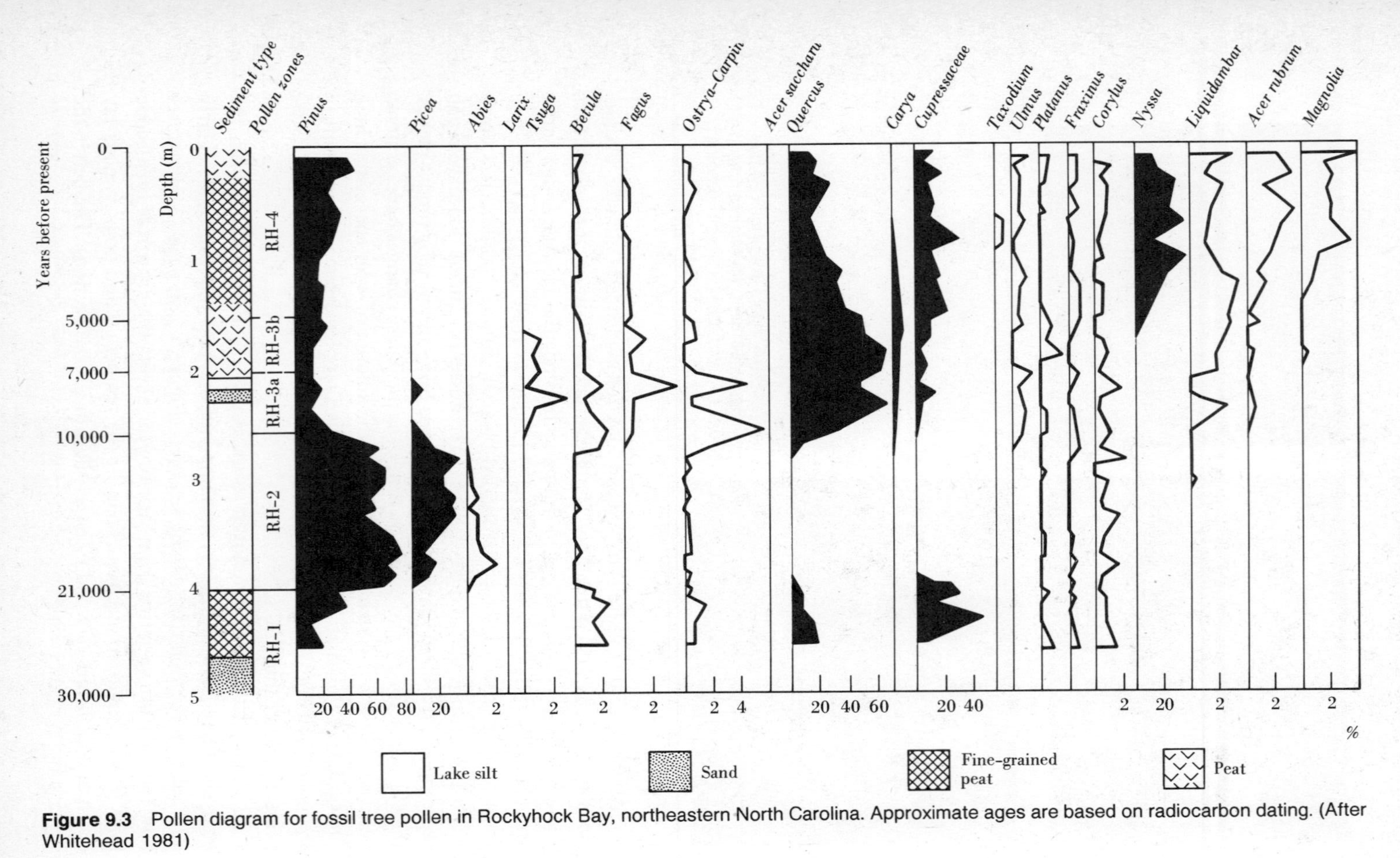

Figure 9.3 Pollen diagram for fossil tree pollen in Rockyhock Bay, northeastern North Carolina. Approximate ages are based on radiocarbon dating. (After Whitehead 1981)

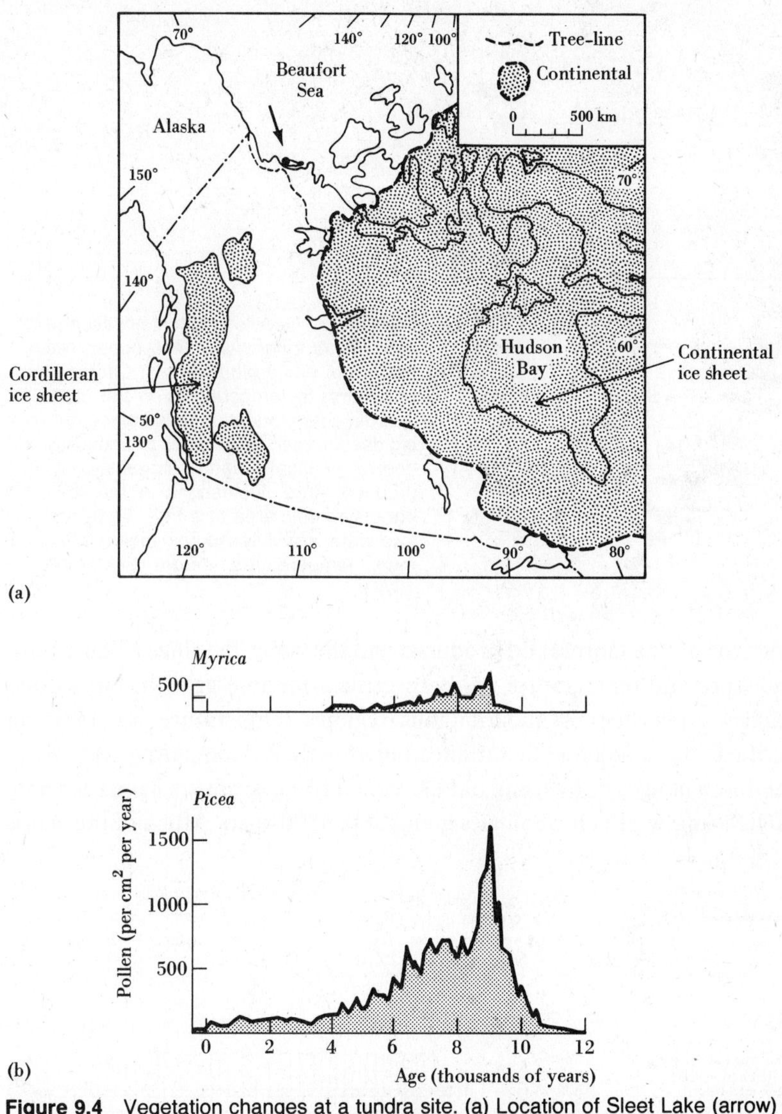

Figure 9.4 Vegetation changes at a tundra site. (a) Location of Sleet Lake (arrow) on the Tuktoyaktuk Peninsula of northwestern Canada. The map shows the approximate position of the large continental ice sheet and the smaller Cordilleran ice sheet 10,000 years ago. Ice sheets began to recede about 14,000 years ago. (b) Pollen deposition rates for two species of boreal forest plants from Sleet Lake. Spruce *(Picea)* forest occupied this site from about 10,000 to 6,000 years ago when the climate was warmer. Bog myrtle *(Myrica)* also occurred there during this period, but is now only found further south. (From Ritchie et al. 1983)

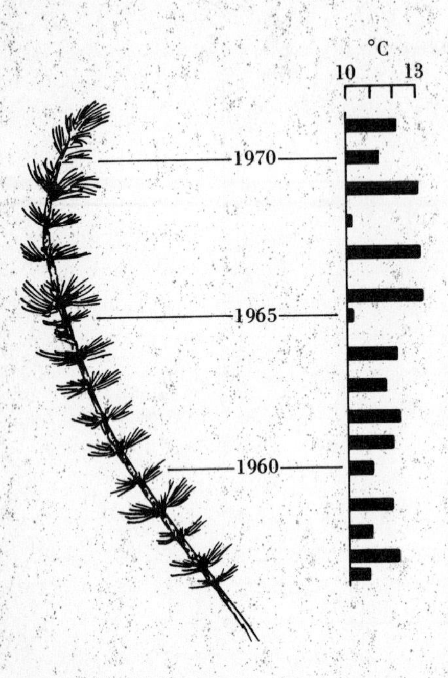

Figure 9.5 Needle length of bristlecone pine *(Pinus longaeva)* at the upper tree line in the White Mountains of California in relation to temperatures in the summer. Needle elongation takes place during the summer. Sequences of unusually cool or unusually warm summers could produce large fluctuations in the total photosynthetic area of a tree. Temperature data are July–August mean maximum temperatures. (From LaMarche 1974)

growing season, the more wood produced and the wider the rings (Figure 9.6). Soil moisture and temperature are both critical for tree growth, but as one approaches timberlines in mountainous regions, temperature becomes the most critical environmental factor affecting growth. By comparing sections of trees of different ages or by using old individual trees, scientists have been able to establish ring-width chronologies going back 5500 years. Much of this work

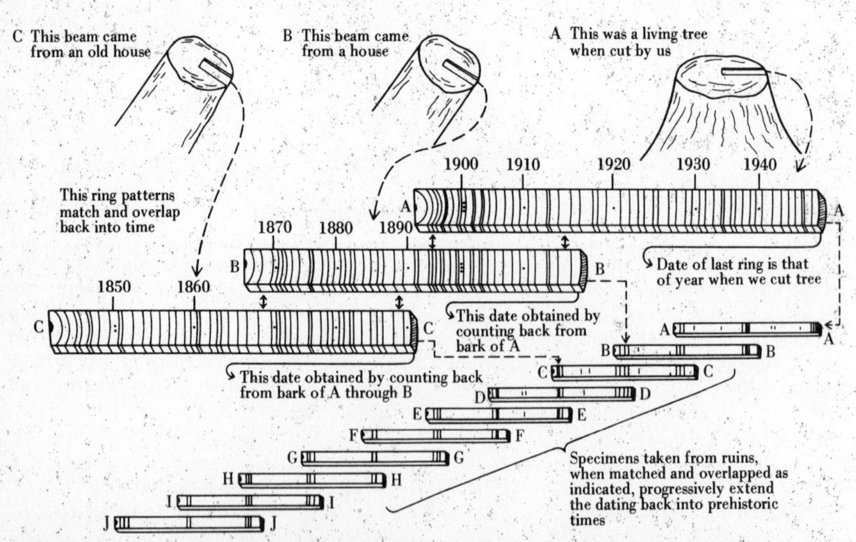

Figure 9.6 Tree-ring analysis. The diagram illustrates how cross-dating is done and how a dated tree-ring chronology is extended backward in time. (From Fritts 1976)

was pioneered at the Laboratory of Tree Ring Research at the University of Arizona founded by A. E. Douglass, who started his work in the early 1900s.

Bristlecone pine has been used extensively in tree-ring research because it grows very slowly near timberline in the mountains of California and thus serves as a good indicator of climate. Individual bristlecone pines may live more than 4600 years and are the oldest living trees. Near timberline, wide tree rings in this pine indicate warm summers and narrow tree rings indicate cool years. Figure 9.7 shows how average ring widths have varied in bristlecone pine during the last 5000 years. Clearly, there is a correlation between poor tree growth and glacier expansion and good tree growth and glacier contraction.

What consequences do these fluctuations in temperature have on animal and plant communities? We can see some of the best evidence on this point by looking at some prehistoric human sites. Stretching across the Great Plains of the United States are the remains of several hundred small Indian villages. They were all abandoned by the time European explorers crossed the Great Plains in the sixteenth century. Indians in these villages planted corn in the summer and hunted during the winter. They were replaced by other Indian tribes that were nomadic hunters. What caused all these villages to be abandoned?

Several sites of ghost villages were excavated in northwestern Iowa in the 1930s and were called by archeologists the Mill Creek culture. This part of Iowa now averages 25 inches (63 centimeters) of rain a year, enough to produce good crops of corn and soybeans. But if a drought occurs and rainfall drops 25 percent, the corn crop fails. Is there any evidence that the Mill Creek culture disappeared because of climatic changes like a prolonged drought?

At the Mill Creek sites only a few inches of soil overlies material left by

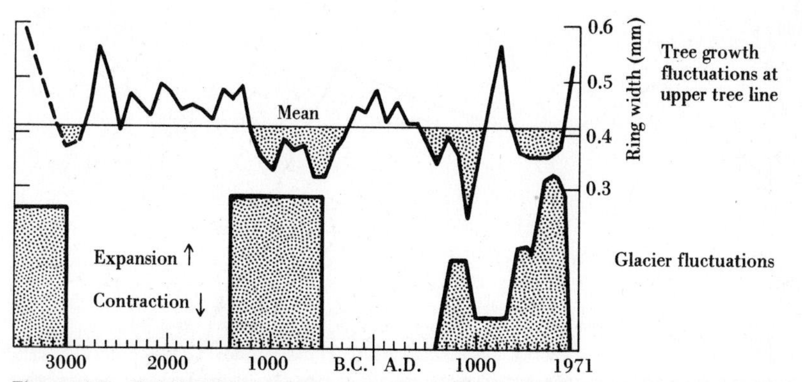

Figure 9.7 Relationship between tree-ring width and glacier fluctuations in the Northern Hemisphere. The upper graph gives the average ring widths of bristlecone pine *(Pinus longaeva)* in the White Mountains of California over the past 5300 years. The lower graph shows periods of expanding and contracting mountain glaciers over the same time period. There is a good correlation between periods of low summer temperatures and periods of growth and advance of glaciers. (After Fritts 1976)

the Indian farmers. Archeologists found most of their informative material in the town dump—bones from animals, broken pottery, ashes from fires, and litter, along with pollen grains that drifted in.

The Mill Creek people ate many types of game animals from fishes and turtles to birds and mammals. Three of the larger animals are of particular interest. Deer and elk are browsing animals and depend on trees for much of their food. Bison are grazing animals characteristic of the short-grass prairies of the western Great Plains. Grasslands are favored by drought and woodlands by abundant rainfall. Consequently, these large mammals are types of climatic indicators. Figure 9.8 shows the proportions of deer, bison, and elk bones found at one Mill Creek excavation site. Older remains are predominately deer.

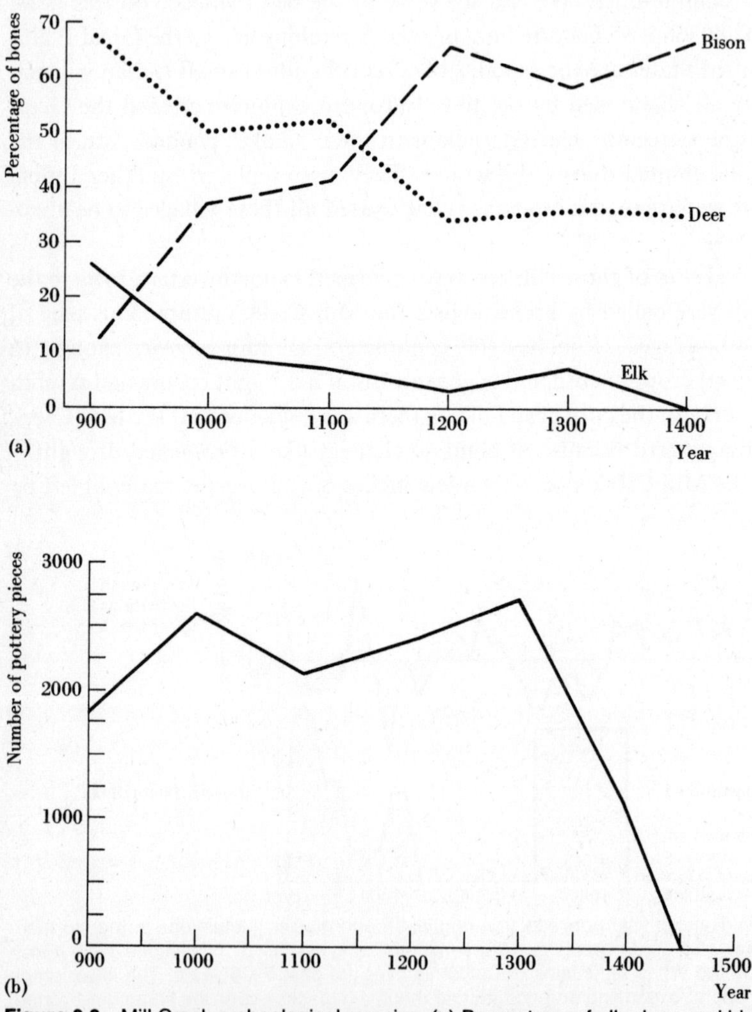

Figure 9.8 Mill Creek archeological remains. (a) Percentage of elk, deer, and bison bones found at one excavation site. (b) Number of pottery pieces found at the same site and used as an index of corn production. (From Bryson and Murray 1977)

After A.D. 1100 the percentage of deer bones dropped and bison came to predominate. But after 1100 the total number of bones also began to decrease, suggesting that hunting was becoming more difficult after the twelfth century.

As an index to corn production archeologists have used the number of pottery fragments in the deposits, since pottery must be used to store and cook corn. Figure 9.8 shows that the number of pottery fragments dropped off rapidly after 1300. By 1400 the Indians had vanished.

All the evidence from Mill Creek is consistent with the idea that drought, beginning about 1200, reduced both corn yield and game abundance and gradually caused these villages to be abandoned. We have here in all probability a thirteenth-century dust bowl.

Pollen samples from the sediments at the Mill Creek sites help to verify this interpretation. Before about 1200 the tree pollen was dominated by oaks; after that time willows came to predominate. Similar changes toward plants characteristic of drier communities occurred in the prairie plants. Aster and sunflower pollen decreased rapidly around 1200 and were replaced by grass pollen from drought-adapted grass species.

There are other suggestions of the collapse of Indian cultures in North America linked with climatic changes. In southern Illinois, just east of St. Louis, stood a major population center of Indians around A.D. 1000. Within a 10-kilometer radius there were many settlements like the one now called Cahokia, which had a population of perhaps 40,000. By the time European explorers arrived, these Indians were all gone. They left an impressive array of mounds. One, Monks Mound, covers an area 300 by 200 meters and is 30 meters high. The labor that went into such mound building shows that these Indians were highly organized and successful.

The Indians who lived at Cahokia grew corn on the floodplain of the Mississippi River. Beginning about A.D. 600 Cahokia grew for 600 to 700 years and then suddenly began to decline after 1200. No one knows why the mound builders of Cahokia disappeared. Perhaps the drought that made Mill Creek uninhabitable also reached into Illinois. Figure 9.9 shows that years of drought in this century follow a geographical pattern consistent with this explanation—a finger of drought reaches over Cahokia.

The Mill Creek farmers and the Cahokia people were not victims of short-term droughts such as we now see in Africa. The changed climate that brought about the collapse of the Indian corn farmers of the Great Plains lasted for about 200 years from 1200 to 1400. After that time the pattern switched back and the rains returned, but by then the people were gone and their culture lost. Both the Mill Creek and the Cahokia settlements were in existence for a time period longer than the United States has existed.

The historical record from Europe and Asia contains many more examples of how humans have been affected by changing climate. Most of these effects have been described for terrestrial ecosystems. Let us look at one recent

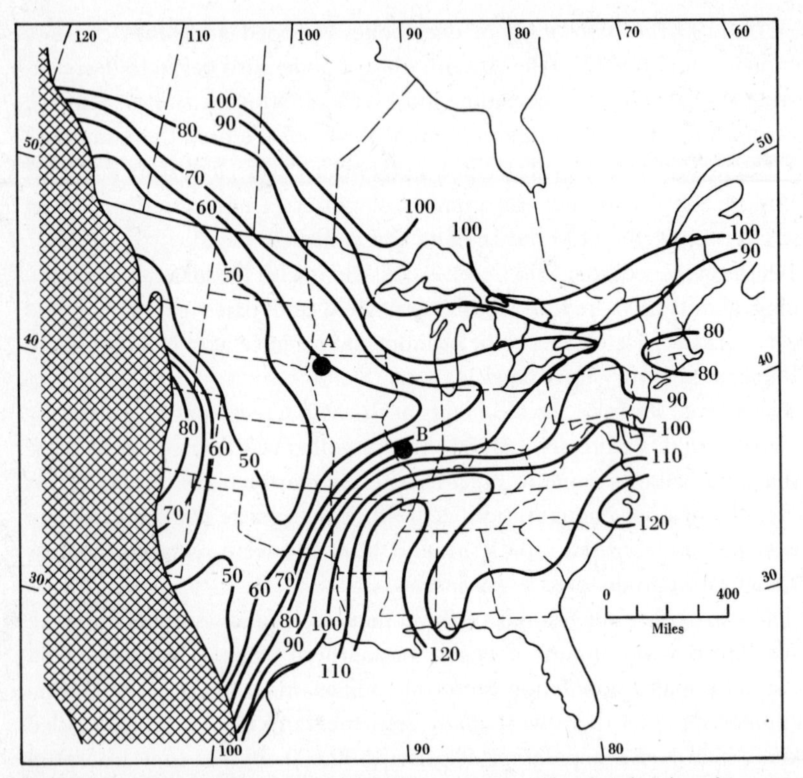

Figure 9.9 Percentage of normal rainfall in average July of major drought years of the 1930s. Note the fingerlike extension of the major drought area into northern Illinois and Indiana to include the Mill Creek sites (A) and the Cahokia sites (B). (Modified from Borchert 1950)

example of a large-scale change in an oceanic ecosystem that has been studied for a long time, the English Channel.

The western part of the English Channel off Plymouth has been studied for more than 75 years and illustrates on a large scale the changes in production and community structure that can occur in aquatic ecosystems. The series of changes that began in the 1920s have now reversed so that a complete cycle can be described. It is called the *Russell cycle* after its principal investigator, the English biologist F. S. Russell.

The complex series of ecosystem changes that make up the Russell cycle can be summarized as follows. A plankton community dominated by the chaetognath *Sagitta elegans* was replaced during the 1930s by one dominated by *Sagitta setosa* (Figure 9.10a on page 140).[1] At the same time there was a

[1]Chaetognaths, or arrow worms, are small invertebrates, 1 to 10 millimeters long, that are one of the larger species in the plankton community of the oceans. They feed on smaller crustaceans in the plankton (see Figure 6.3).

drop in the dissolved phosphate concentration of the seawater. Phosphate is a critical nutrient for phytoplankton in the spring. The abundance of zooplankton dropped at the same time, and the numbers of young fish in the plankton fell dramatically (Figure 9.10c). Herring, which feed on plankton, decreased, and pilchard replaced herring as the dominant pelagic fish in the English Channel (Figure 9.10b). During the 1970s the cycle reversed. *Sagitta elegans* returned to the plankton, pilchard declined, and larval fish became abundant in the plankton. The net result was that by 1980 the English Channel had returned to a stage approximately like that of the 1920s.

What caused the Russell cycle? Three major hypotheses have been suggested to explain these ecosystem changes: (1) nutrient control, (2) competition, and (3) climatic changes.

The nutrient control hypothesis explains these changes by variations in the flow of oceanic water into the English Channel from the west. Oceanic water typically contains less of the critical inorganic nutrients like phosphate that limit primary production. This theory has been rejected because no change in primary production was measured during the reversal of the cycle from 1964 to 1974, when phosphate levels increased. Also, variations in oceanic currents are usually produced by climatic changes, and any correlation with nutrient changes may thus be secondary effects of climate.

The competition hypothesis states that the community in the English Channel can exist in two alternate stable states—a *Sagitta elegans*–herring community and a *Sagitta setosa*–pilchard community. Pilchard were able to exclude herring and assume dominance in the 1930s, possibly aided by overfishing on herring, according to this hypothesis. But pilchard have now been replaced by mackerel, and herring are still scarce, and the decline of the pilchard cannot be due to overfishing. The competition hypothesis does not explain the changes shown in Figure 9.10 convincingly.

The climatic change hypothesis is generally agreed to be the underlying explanation of the Russell cycle. Figure 9.11 on page 141 shows the average change in sea surface temperature in the English Channel since 1924. A distinct warming and then cooling trend is seen, which correlates with a general pattern of climatic fluctuation that has been measured in the Northern Hemisphere (see Figure 9.1). Rises and falls in sea temperature may result in sudden switches in the ecosystem at a faunal boundary. Herring, for example, is a cold-water fish near the southern edge of its distribution in the English Channel. Pilchard is a warm-temperate species close to its northern limit. The changes over the Russell cycle are broadly consistent with a temperature control hypothesis, and the apparent stability of marine ecosystems over a short span of years may in fact be part of a long cycle of dynamic change.

One other important aspect of the Russell cycle is the apparent disparity

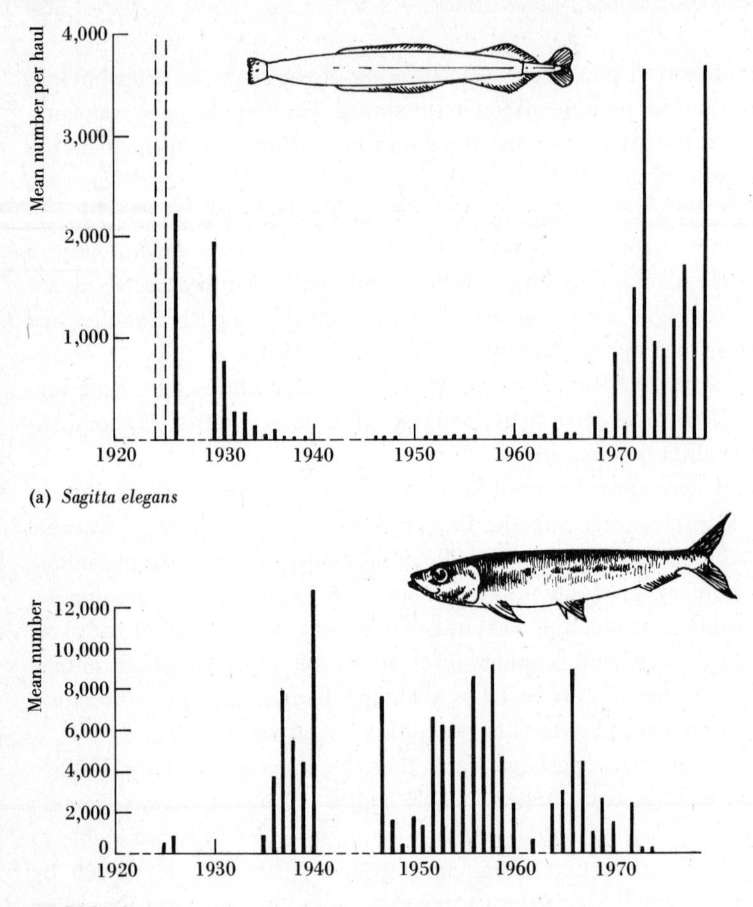

(a) *Sagitta elegans*

(b) Pilchard eggs

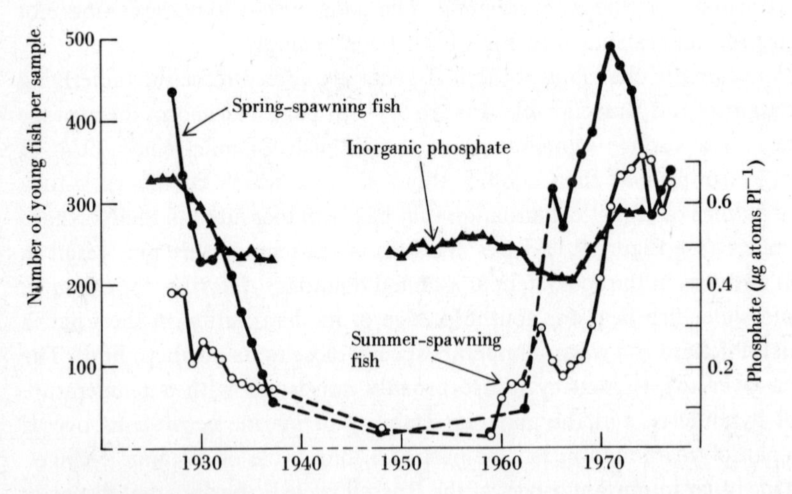

(c) Young fish and phosphate

Figure 9.10 Some of the biological changes in the Russell cycle of the English Channel. These data, taken off Plymouth, are based on weekly sampling near the Eddystone reef. (a) Numbers of the chaetognath *Sagitta elegans* in 5000 cubic meters of water during the summer months. (b) Number of pilchard *(Sardina pilchardus)* eggs per 2-meter net haul during the spawning season. The number of eggs is assumed to be an index of the number of adult pilchards. (c) Mean amount of inorganic phosphate in winter and mean number of small fish in a 2-meter net sample. The small fish are divided into spring spawners and summer spawners. (After Southward 1980)

between the biological fluctuations and the physical-chemical changes. Sea temperature has changed only 0.5°C, and phosphate less than 25 percent. But this has resulted in the virtual disappearance of some zooplankton and fish species and a 10-fold change in populations of others. The impact of climatic change on ecosystems cannot be judged by the size of the temperature shift. Small changes may produce enormous effects in ecosystem structure and function.

It is important to remember at this point that not all the changes that occur in biological communities are due to climatic changes. For example, since detailed observations began around 1850, many birds have extended their geographical ranges in northern Europe. The bird communities in southern Finland, for example, contain more species now than they did in the last century. Most of this colonization is not due to climatic amelioration, but to habitat changes resulting from agriculture and forestry. If there has been a climatic component to bird distributional changes during the last few years, it has been dwarfed by the effects that humans have had on the landscape.

One of the difficulties of thinking about climatic changes is that climatic trends show up slowly and we cannot predict the future well. Figure 9.12

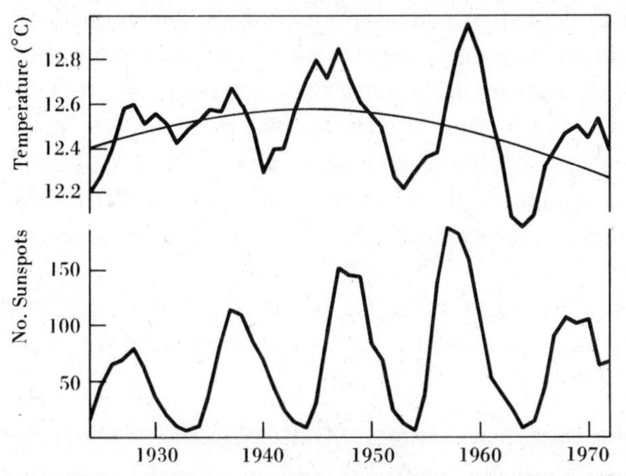

Figure 9.11 Average sea surface temperature in the English Channel compared with mean annual sunspot numbers. The 11-year cycle in sunspot numbers is reflected in sea surface temperatures, but superimposed on this is a global trend in temperature shown by the smooth curve. (After Southward et al. 1975)

shows generalized temperature curves for the Northern Hemisphere, and helps to put the present climate in perspective. Most of the last 150,000 years has been occupied by a glacial period, and to find a time as warm as the last few years, we have to go back about 125,000 years to the last interglacial period.

After the end of the last ice age, about 15,000 years ago, global temperatures began to rise rapidly and during the last 10,000 years they have fluctuated irregularly. We are in an interglacial period. In the past interglacials have occurred about once in every 100,000 years and each has lasted 8,000 to 12,000 years. The present interglacial has already gone on for about 11,000 years.

One thing about climate is certain—the climate we think of as "normal" and the weather office calls "normal" is not typical of the last 1000 years and is much less typical of the last million years. Since 1945 the Northern Hemisphere has cooled about 0.6°C (1°F). Cooling could continue, or temperatures could reverse again. It is impossible to predict which.

Why do climates change? The sun generates our climate, and one possible cause of climatic change could be a change in the output from the sun. The sun does fluctuate in output, as the 11-year sunspot cycle shows (see Figure 9.11), but a changing sun does not seem to be the main cause of climatic changes. A major cause of climatic change is the way the earth's orbit shifts around the sun. The average distance of the earth from the sun varies somewhat, as does the shape of the earth's orbit. And the earth also wobbles slightly on its axis. These orbit variations change regularly in cycles over thousands of years; for example, one complete wobble of the earth's tilt takes 40,000 years. A Yugoslav mathematician, M. Milankovitch, calculated these variations in 1920 and produced a model that correlated the past ice ages with the earth's orbit changes. His calculations match up well with the general pattern of the ice ages, and if his predictions are correct, a new ice age will begin within a few thousand years. These changes in the earth's position relative to the sun are now believed to be the major cause of climatic change.

But there are other ways of changing climate as well. Anything in the air that can affect the net amount of the sun's energy that the earth receives is a potential source for climatic change. Two substances are of major concern, *dust* and *carbon dioxide.*

The concentration of carbon dioxide has increased about 10 percent during the last 80 years because of human activity—burning coal, oil, and

Figure 9.12 Northern Hemisphere temperatures for the past 150,000 years with more detail for recent years. (a) The last 150,000 years as indicated by marine deposits, pollen records, and shoreline changes. (b) The last 25,000 years as indicated by tree rings, pollen records, and glacier records. (c) The last 1,000 years. a, Mill Creek cooling; b, Little Ice Age; c, early twentieth-century warming. (d) The last 100 years, showing the cooling that has occurred since 1950. (From Bryson and Murray 1977)

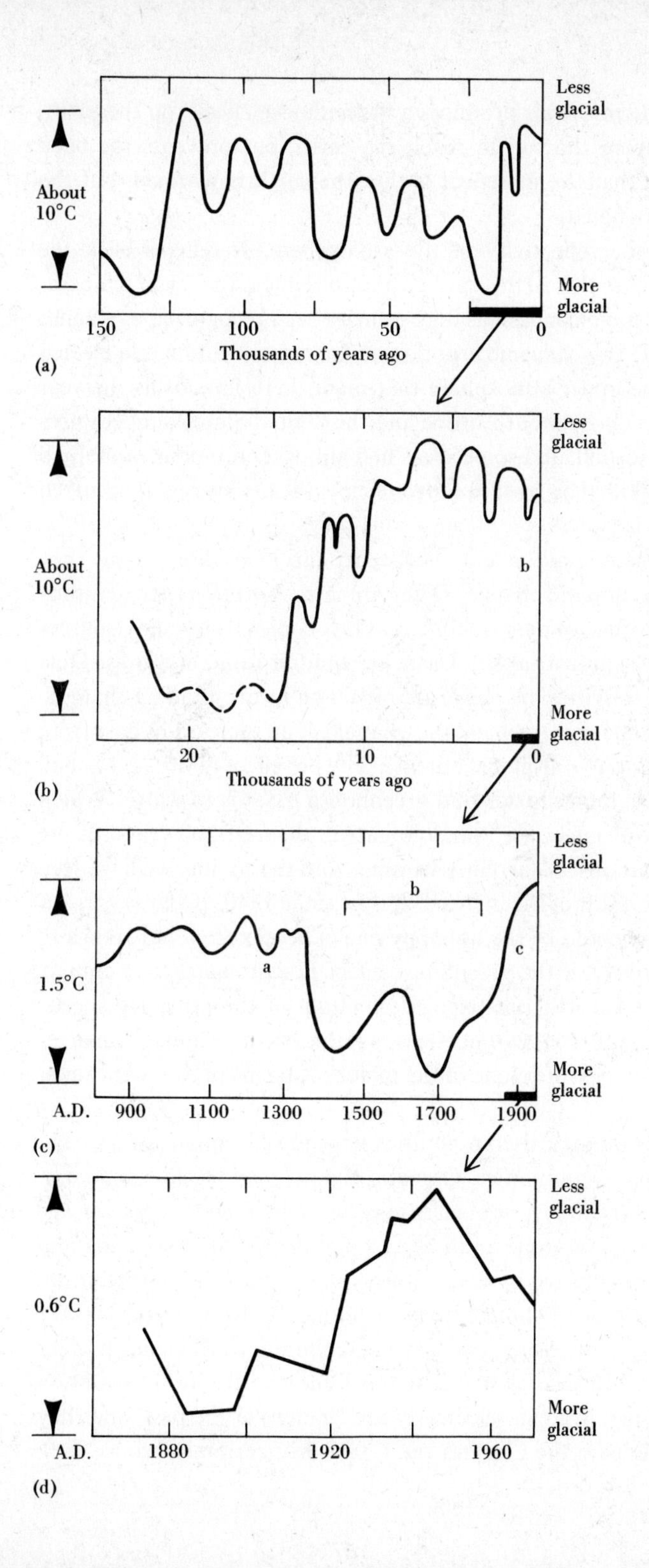
Less glacial
About 10°C
More glacial
150
100
50
0
Thousands of years ago
(a)
Less glacial
About 10°C
b
More glacial
20
10
0
Thousands of years ago
(b)
Less glacial
b
a
c
1.5°C
More glacial
A.D.
900
1100
1300
1500
1700
1900
(c)
Less glacial
0.6°C
More glacial
A.D.
1880
1920
1960
(d)

wood. Carbon dioxide in the air produces a "greenhouse effect" on the earth. It allows radiation from the sun to reach the earth, but prevents the back radiation of heat, so that the net effect is that the earth is warmed, but the upper atmosphere is cooled.

Dust changes the reflectivity of the atmosphere. It reflects back the sunlight before it reaches the earth's surface and so reduces the mean temperature. Dust ejected from volcanoes has for centuries been recognized as a cause of cold climates. In 1815 a volcanic eruption at Tambora in Indonesia ejected a veil of dust into the upper atmosphere that made 1816 famous as the year that had no summer. Throughout Europe and the United States temperatures averaged 1°C below normal, and some areas had almost continuous rainfall all summer. Crops were killed by frost or failed to ripen. Riots over food occurred in Wales and Ireland.

Volcanic eruptions may have a short-term effect on climate, but they cannot explain major climatic changes. They are important to us because they alert us to the significance of dust to climate. Dust comes from many sources other than volcanoes—industrial smokestacks, wind erosion, and fires. Dust generated by human activities may have an impact on future climatic changes. But there is considerable disagreement on what effect an increase in dust from human activities can do to our global climate. The net effect of dust is to cool, and the net effect of carbon dioxide and greenhouse gases is to warm. Which of these two contrasting effects will predominate in the next 200 years? There is some suggestion that dust is currently winning, and the cooling we have seen since 1950 is a direct result of the increase in dust since 1940. If this is correct, our impact on climate could be the unhappy one of accelerating our return to the ice age. Alternatively, if the greenhouse effect predominates, our climate will warm, and sea level will rise from the melting of the polar ice sheets. Coastal cities would be affected within 50 years if this occurs. Climatic changes caused by human activity will be one of the major problems of the twenty-first century.

To summarize: Biological communities depend on climate, and as climate changes, so must communities. Climatic change may be slow and communities may change gradually, or change may be rapid. Since the last ice age ended, temperate and polar zone animals and plants are still colonizing and recovering in the areas affected by glaciation. Animal and plant communities are never in equilibrium with climate because climate is always changing. We should therefore not view either our present communities or our present climates as constant and everlasting. They are but one frame in a motion picture that never slows. We can see clearly the changes of the past, and they must help us to anticipate the changes the future will certainly bring.

FURTHER READING

Birks, H. J. B., and H. H. Birks. 1980. *Quaternary Palaeoecology.* Edward Arnold, London.

*Bryson, R. A., and T. J. Murray. 1977. *Climates of Hunger: Mankind and the World's Changing Weather.* University of Wisconsin Press, Madison.

Ford, M. J. 1982. *The Changing Climate: Responses of the Natural Fauna and Flora.* Allen and Unwin, London.

Fritts, H. C. 1976. *Tree Rings and Climate.* Academic Press, New York.

Godwin, H. 1981. *The Archives of the Peat Bogs.* Cambridge University Press, London.

Lamb, H. H. 1982. *Climate, History and the Modern World.* Methuen, London.

Schneider, S. H., and R. Londer. 1984. *The Coevolution of Climate and Life.* Sierra Club Books, San Francisco.

*Highly recommended.

chapter 10

Natural Systems Are Products of Evolution

Communities of plants and animals have a history, and we call that history *evolution.* We begin this chapter with the observation that evolution has occurred in the past to give us the organisms we see today. How has this occurred? Can we use our knowledge of how modern communities evolved to help us to design better systems of forestry and agriculture? How much are organisms constrained by past evolutionary events? These are all important questions at the interface between ecology and evolution. In this chapter I will sketch the answers that are slowly emerging to these questions.

Evolution is descent with modification, as Darwin put it, or the cumulative change in the characteristics of organisms over many generations. Evolution was recognized for at least a hundred years before Charles Darwin, whose genius was to recognize that the mechanism for evolution is *natural selection.* Natural selection is a process that arises from three biological observations: (1) variation among individuals in some attribute (such as eye color or body size), (2) consistent differences in reproduction or survival among individuals having this attribute, and (3) inheritance, or a resemblance between parents and offspring for the attribute. The process of natural selection operates to produce organisms that are adapted to the environments in which they live. Favorable attributes are those that give higher reproductive rates or better survival. Individuals with such favorable attributes are "selected for," and thus these attributes become more common in the population over several generations. But natural selection does not produce one "perfect" model for each species.

Biologists have discovered that there is a great amount of genetic variation existing *within* each species.

One of the greatest and persistent errors in biological thought is to think of each species as a constant and fixed type in which all individuals are genetically equal. Darwin recognized this mistake, but the full implications of Darwin's ideas were not appreciated until this century. In the 1920s a Swedish botanist, Göte Turesson, began to describe local adaptations within plant species. His basic technique was to collect plants from a variety of areas and grow them together in field plots at one location. The type of result he obtained in this early work can be illustrated with one example. *Plantago maritima* grows as a tall, robust plant in marshes along the coast of Sweden and also as a dwarf plant on exposed sea cliffs in the Faeroe Islands. When plants are grown side by side in an experimental garden, this height difference is not as extreme but remains significant.

Source of *Plantago maritima*	Mean height in experimental garden (cm)
Marsh population	31.5
Cliff population	20.7

Turesson called these *ecotypes*—genetic varieties within a single species. These varieties represent adaptations to the local environment.

This transplant technique is an attempt to separate the phenotypic and genotypic components of variation. Plants of the same species growing in such diverse environments as sea cliffs and marshes can be different in morphology and physiology in three ways: (1) All differences are phenotypic (that is, observed differences are caused by environmental conditions in which the plants grow), and if seeds are transplanted from one situation to the other, they will respond exactly as the resident individuals. (2) All differences are genotypic (that is, observed differences are caused by genetic variations), and if seeds are transplanted from one situation to the other, the mature plants will retain the form and the physiology typical of their original habitat. (3) Some combination of phenotypic and genotypic determination produces an intermediate result. In natural situations the third case is most usual. Many examples are now described, particularly in plants.

One of the most intensively studied set of ecotypic races occurs in the perennial herb *Achillea* (yarrow), analyzed by J. Clausen and his colleagues working at Stanford in California. Three very similar species of *Achillea* occur, two from western North America and one from Europe. Clausen and his colleagues studied the two North American species in detail.

Figure 10.1 shows the genetic variation that can occur between latitudi-

(a)

(b)

Figure 10.1 Ecotypic variation in *Achillea borealis* as grown in a Stanford, California, experimental garden. A race from Selma, California (a) and another from Seward, Alaska (b) are shown reproduced to the same scale. The approximate temperature ranges of their native habitats are given in (c). (After Clausen et al. 1948)

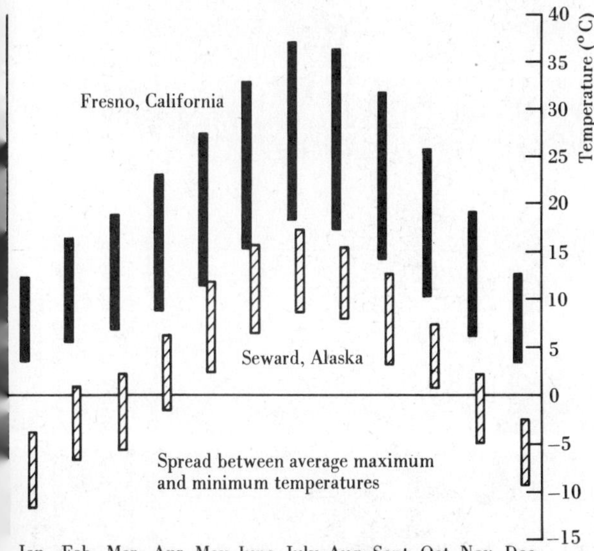

(c)

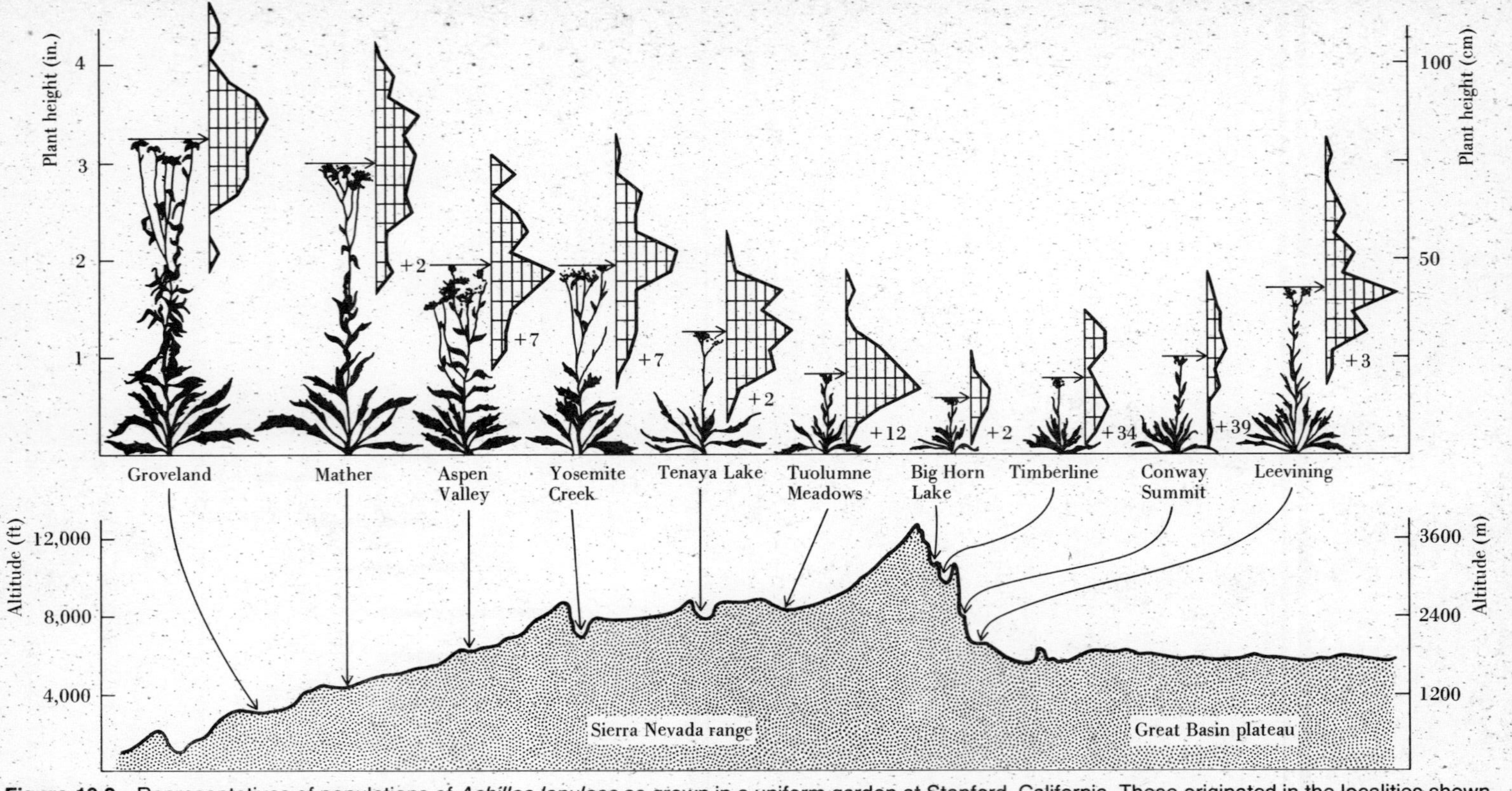

Figure 10.2 Representatives of populations of *Achillea lanulosa* as grown in a uniform garden at Stanford, California. These originated in the localities shown in the profile of a transect across central California at approximately 38 degrees north latitude. Altitudes are to scale; horizontal distances are not to scale. The plants are herbarium specimens, each representing a population of approximately 60 individuals. The frequency diagrams show variation in height within each population; the horizontal lines separate class intervals of 5 centimeters according to the marginal scale; and the distance between vertical lines represents two individuals. The numbers to the right of some frequency diagrams indicate the nonflowering plants. The specimens represent plants of average height, and the arrows point to mean heights. (After Clausen et al. 1948)

nal races of *Achillea borealis* from California and Alaska. Adaptation to the local environment has produced plants so different we might think they are different species.

In the Sierra Nevada of California races of *Achillea lanulosa* occur. As one proceeds up these mountains, the average winter temperature decreases below freezing, so winter dormancy is necessary and plants are smaller. On the eastern slope of the Sierra Nevada plants of *A. lanulosa* are late-flowering and are adapted to the cold, arid climate. Clausen and his colleagues collected seeds from a series of populations of *A. lanulosa* across California and raised plants in a greenhouse at Stanford, with the results shown in Figure 10.2. The major attributes of these races are maintained when grown under uniform conditions in the same place.

Genetic variation between local races can be economically important in forestry. Sitka spruce *(Picea sitchensis)* is a conifer native to the west coast of North America. It has been planted widely in Britain. Young sitka spruce planted in upland areas of Britain frequently suffer needle damage following unseasonal frosts in late spring and early autumn. Needle damage in autumn has occurred in Scotland in 1971, 1972, 1974, 1976, 1979, and 1983. Damage was greatest on spruce grown from seed of southern origin (Oregon and Washington) and least on spruce grown from Alaskan seed.

Figure 10.3 shows the observed probabilities of autumn frost from a series of sites in western North America, where sitka spruce is native, and a series of British sites. The autumn frost climate is similar in Edinburgh and in coastal Oregon (Cloverdale), in Durham and Vancouver, and in Kielder Castle, Eskdalemuir, and other upland plantations in Scotland and Sitka and Cordova in coastal Alaska. Temperate tree species, like sitka spruce, become more frost hardy in the autumn as temperatures cool and day lengths shorten. The degree of cooling and the day lengths needed to induce frost hardening are fixed by natural selection in relation to the environments where the spruce originated.

The pattern of frost hardening in autumn is shown for three races of sitka spruce in Figure 10.4 on page 153. The three races clearly respond differently to the declining temperatures and shortening day lengths of the site at which they were all growing. Most of the sitka spruce planted in Britain have come from the Masset area of British Columbia. Masset trees are susceptible to severe frosts in upland Scottish sites, and if the climate does not change, one could predict frost damage to sitka spruce at these upland sites about once every 8 to 11 years. A similar problem occurs for spring frost damage, which can also be severe in upland plantations in Scotland.

The moral is simple: *Do not assume that a tree is a tree is a tree!* Because natural selection produces adaptation to the local environment, every species is a montage of different subpopulations genetically adapted on a local scale.

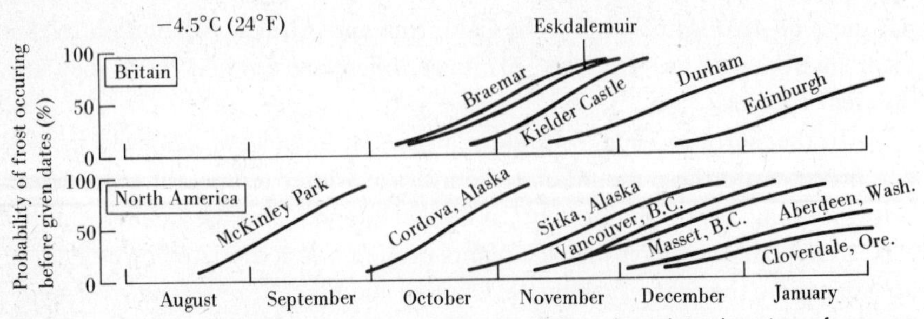

Figure 10.3 Probabilities of severe frosts occurring before given dates in autumn for seven sites in North America (spanning the geographical range of sitka spruce) and for five sites in Britain where sitka spruce has been planted. (From Cannell 1985)

As forestry becomes more like agriculture, the geographical origin of the trees being planted becomes critical to success.

Not all adaptation is concerned with climatic factors and I want to discuss next the impact of biological relationships on adaptation. We are particularly interested here in the interactions that occur *between* species and how these might affect the structure of biological communities.

Six general categories of interactions can be recognized between two species. We can characterize them most simply with the symbols + for beneficial effect on the species, — for adverse effect on the species, and 0 for no effect.

	Species A	Species B
Competition	—	—
Predation or parasitism	+	—
Mutualism	+	+
Commensalism	+	0
Amensalism or inhibition	—	0
Neutralism	0	0

In all these interactions except neutralism there is incipient evolutionary interplay between species and potential evolutionary pressures for readjustment.

Competition between species is widespread in natural communities and may be one important interaction affecting the structure of communities. Competition occurs over resources that are in short supply and, while any resource may be an important focus of competition, food is most often the battleground of competition. The English ornithologist, David Lack, was one of the first to recognize that closely related bird species, which one might expect to compete for food, usually either eat different foods or live in different habitats so that competition is minimized. Figure 10.5 on page 154 illustrates how natural selection may operate to minimize competition for food between species that eat the same things.

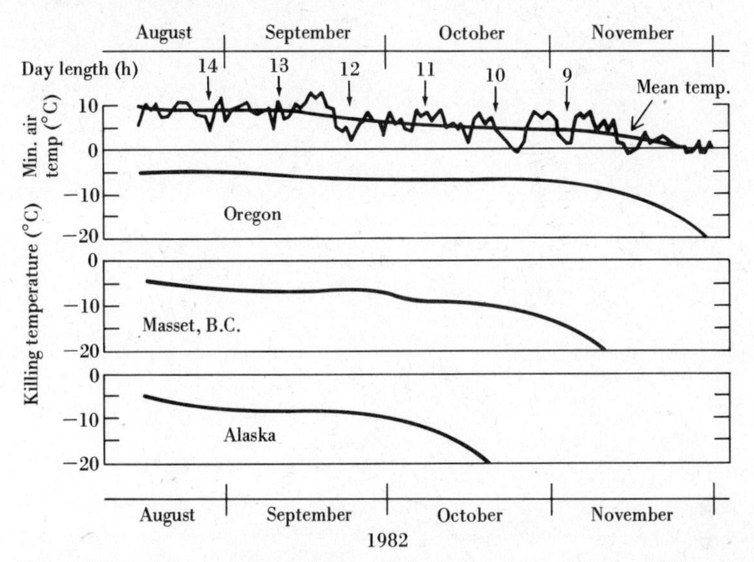

Figure 10.4 Autumn frost hardening of shoots from three races of Sitka spruce grown in a common nursery in Scotland. Upper graph gives daily minimum air temperatures for 1982 and day lengths. The lower three graphs show the observed shoot-killing temperatures for spruce trees originating from southern populations (Oregon), central populations (Masset), and northern populations (Alaska). Frosts that will kill the Oregon race in October and November will not affect the Alaskan race. (From Cannell et al. 1985)

One evolutionary consequence of competition between two species has been the divergence of the species in areas where they occur together. This type of divergence is called *character displacement* and is illustrated in Figure 10.6 on page 155. A classic example of character displacement occurs in Darwin's finches on the Galapagos Islands off Ecuador (Figure 10.7 on page 156). In this example beak depth is used as an index of food size taken. On three sets of islands *Geospiza fuliginosa* and *G. fortis* occur together, and their beak sizes do not overlap. On two islands *G. fortis* and *G. fuliginosa* occur by themselves, and, in the absence of competition, their beak sizes are similar.

What we see today is a result of evolution in the past, and the structure of natural communities, as in these Darwin's finches, may be partly the result of competition in the past. These effects have been described as the "ghost of competition past" because modern communities may show little sign of present-day competition even though they evolved in the past because of competitive pressures for food. If competition is too intense between species, it may result in the extinction of one of the competitors very quickly.

Interactions between predators and prey have resulted in much evolution in natural communities. Some of the best examples come from animals that show warning coloration and are involved in *mimicry* complexes. Mimicry is the resemblance of one species to another, usually for protective purposes.

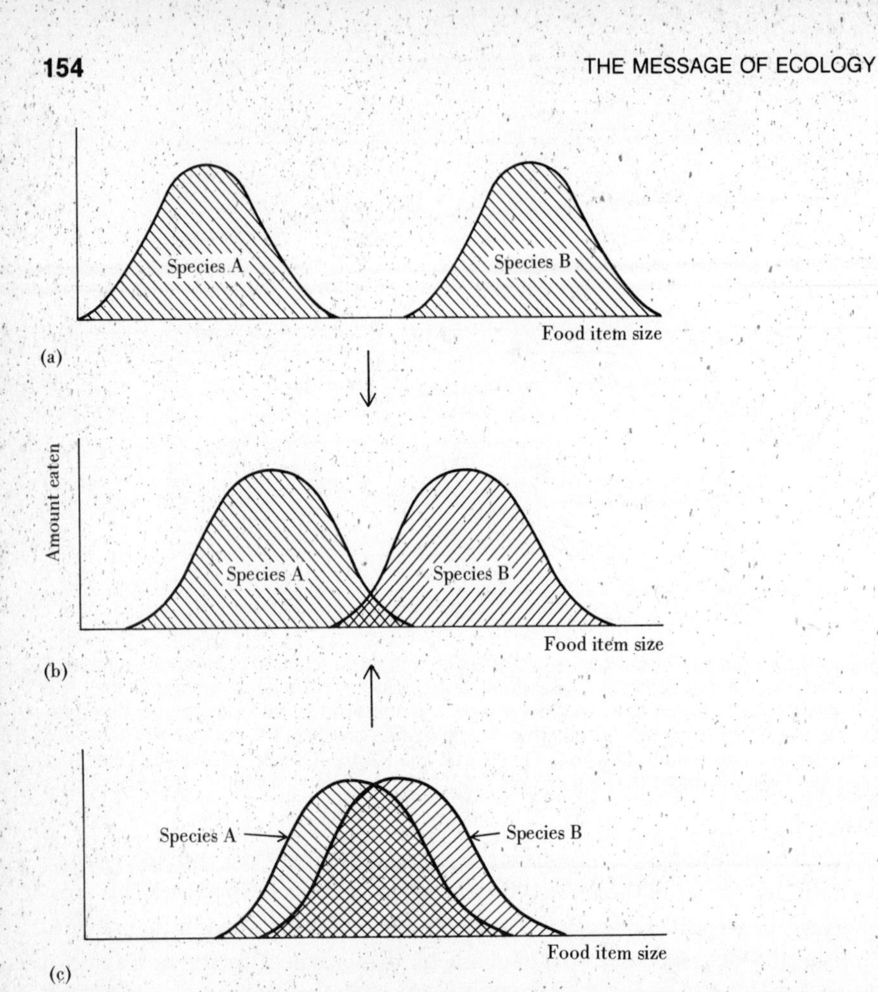

Figure 10.5 Hypothetical food utilization curves for two species. Food size is the resource for which competition may occur in this hypothetical case. Arrows indicate the direction of evolutionary pressures toward case (b). In case (a) there is unutilized food between the two species. Natural selection will favor any individual who feeds in this unutilized zone, and the system will evolve toward case (b). In case (c) there is intensive competition and natural selection will favor individuals who feed on smaller items for species A and larger items for species B, so the system will evolve toward case (b). (From Krebs 1985)

Mimicry falls into two main categories. *Batesian* mimicry[1] involves species playing two different roles: (1) *models,* which are individuals of the species that are poisonous or can sting, and (2) *mimics,* which are individuals that lack poisons or protection and gain their advantage by resembling the models. By contrast, *Mullerian* mimicry[2] involves several species, all poisonous or dangerous, that resemble one another and thereby gain additional protection from predators.

[1]Named after the naturalist H. W. Bates who described resemblances among South American butterflies in 1862.

[2]Named after the naturalist Fritz Müller who described the evolution of protective similarity in 1879.

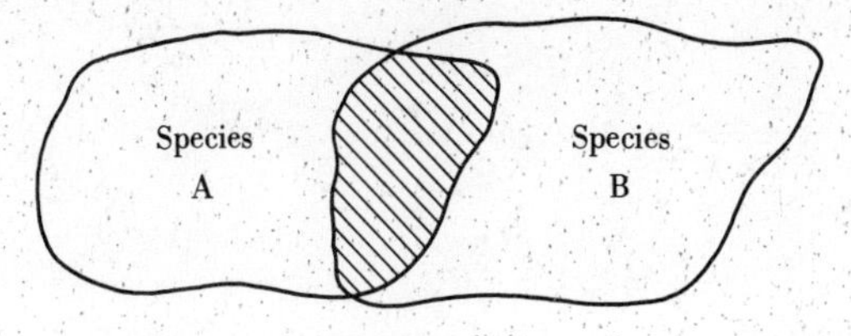

(a) Geographical distribution

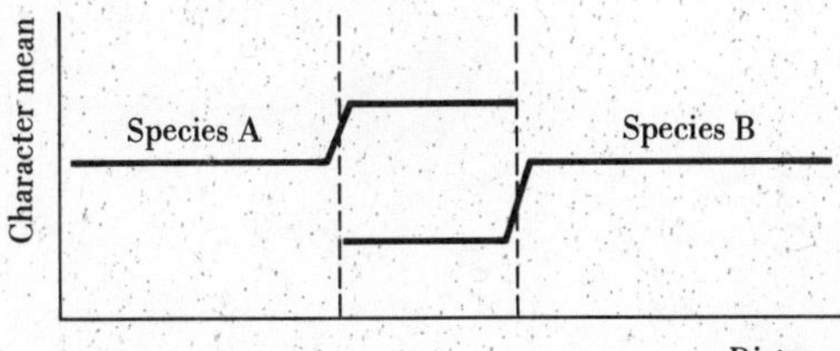

(b) Character changes

Figure 10.6 Schematic view of character displacement arising from competition between two species in the zone of overlap. This scheme is suggested as an explanation of the observations in Figure 10.7. (From Krebs 1985.)

Butterfly mimicry has been particularly well studied because of the interest of butterfly collectors in color variations and the great range of genetic variation in butterfly color. One of the classical mimicry systems in North America is that of the monarch butterfly *(Danaus plexippus)* and the viceroy butterfly *(Limenitis archippus)* (Figure 10.8 on page 157). The monarch is the model in this system and it becomes poisonous in a clever way.

A milkweed *(Asclepias curassavica)* grows abundantly in Costa Rica and other areas in Central America, but cattle will not eat it. This milkweed contains secondary plant substances called cardiac glycosides that affect the vertebrate heartbeat and are poisonous to mammals and birds. But certain insects are able to eat milkweeds without harmful effects, and among these are the danaid butterflies, including the monarch and queen butterflies. The danaid butterflies are known to be distasteful to insect-eating birds and serve as models in several mimicry complexes. Monarch butterflies have developed biochemical mechanisms for feeding on milkweeds containing cardiac glycosides and then storing these poisons in their tissues so that the insects acquire chemical protection from the plants they eat.

To test this hypothesis, monarch butterflies were raised on cabbage (which contains no cardiac glycosides) and found to be completely acceptable to bird predators. However, birds that fed on monarch butterflies raised on the milkweed *A. curassavica* became violently ill within 12 minutes, vomited the insects, and then recovered within 30 minutes. Such birds learned quickly to reject all monarch butterflies on sight. This rapid learning by birds allows the monarch to trick its predators, because not all milkweeds contain cardiac

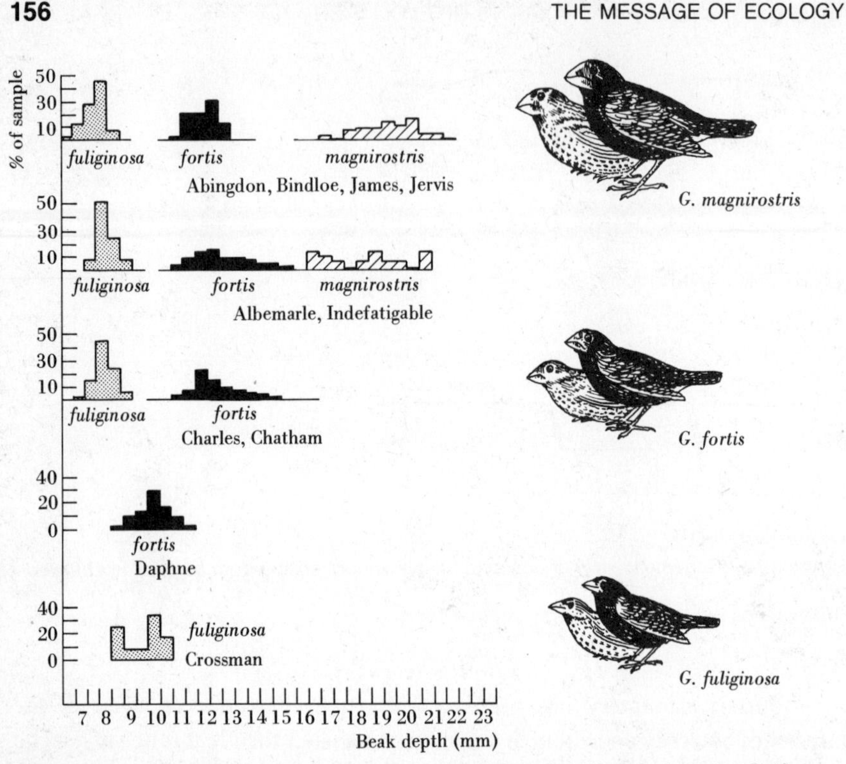

Figure 10.7 Character displacement in beak size in Darwin's finches from the Galápagos Islands. Beak depths are given for *Geospiza fortis* and *G. fuliginosa* on islands where these two species occur together (upper three sets of islands) and alone (lower two islands). *Geospiza magnirostris* is another large finch that occurs on some islands. (After Lack 1947)

glycosides. For example, three of the common milkweed species of eastern North America are nontoxic, and monarchs raised on these milkweeds are edible. However, if a vertebrate predator learns to avoid eating monarch butterflies after one unpleasant experience, the edible monarchs escape predation because they look exactly like toxic monarchs. In the same way edible viceroy butterflies escape predation by looking like toxic monarch butterflies. Bird predators can only be "fooled" by this mimicry system if the frequency of viceroys is not too high. Naive young birds that fed only on viceroy butterflies would conclude that all butterflies of these colors are edible, and the system would collapse.

Systems of predators and prey may be subject to an "arms race" in evolutionary time. If foxes are selected to be more clever at catching rabbits, rabbits must evolve to become more wary of foxes, and so *ad infinitum.* Arms races may lead to the eventual extinction of either the prey or the predator or

Figure 10.8 Batesian mimicry in the monarch *(Danaus plexippus)* and viceroy *(Limenitis archippus)* butterflies of North America. The viceroy mimics the poisonous monarch which is the model species for this system. Note how far the Viceroy has changed from its close relatives in order to mimic the Monarch butterfly. (Photo courtesy of L. P. Brower)

both, but this is unlikely to occur unless the predators feed on only one species of prey.

Large cursorial (running) predators and their ungulate prey are an ideal system to analyze for an arms race style of coevolution. Ungulates flee from predators by running at high speed. Running speed is a function of limb form and length, and an abundant fossil record allows us to look for evolutionary trends for the last 60 million years (when mammals first appeared in the fossil record). We can construct an index of running speed by measuring the elongation of the long bones of the foot in both ungulate prey and their predators. Figure 10.9 gives the results. Ungulates have evolved to run faster and faster over the last 60 million years, as have their predators. But surprisingly, there is an ever-widening gap between prey and predators. This evidence suggests that mammalian predators were not evolving fast enough to keep up with their prey, that predators were losing in the arms race. In some cases extinction might have resulted from this widening gap.

Some systems that appear to be a case of predation are found to be something else upon closer investigation. Seed eating provides some good examples. Plant seeds appear to be "prey," and birds and mammals appear to be "predators," but many plants depend on birds and mammals for the dispersal of their seeds. Reciprocal evolution between plants and their seed predators may help to explain some features of plant reproductive biology

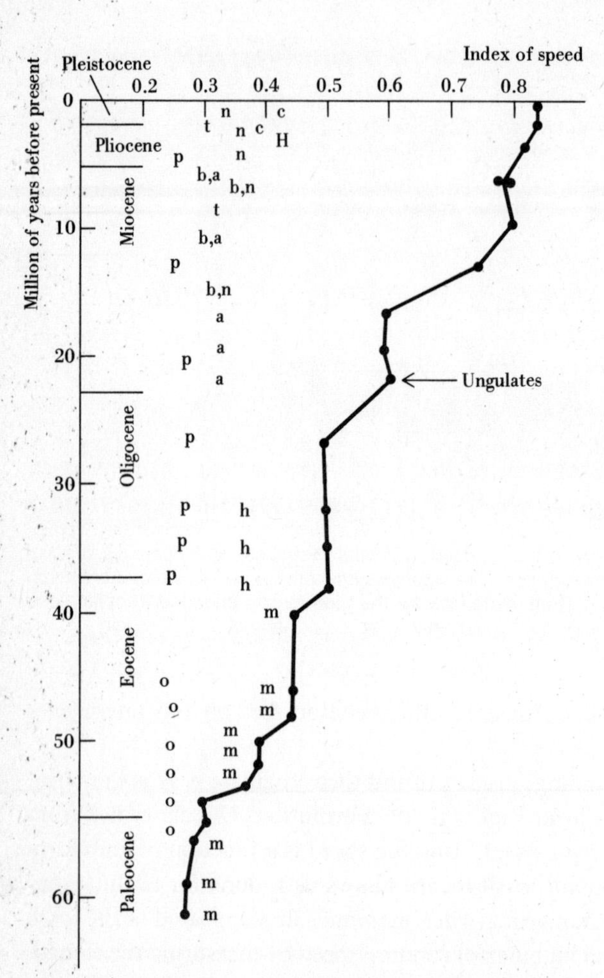

Figure 10.9 Index of speed in running for North American ungulates and their predators. Letters indicate taxonomic subgroups of predators. Solid line connects ungulate data. Ungulates appear to be winning in this arms race. (From Bakker 1983)

and animal feeding habits. Both the "prey" and the "predators" may gain in these relationships. This may be seen particularly clearly in the evolution of fruits.

Fruits are basically discrete packages of seeds with a certain amount of nutritive material. Vertebrates eat these fruits and digest part of the material thereby gaining food energy, but many of the seeds pass through the digestive system unharmed. The resulting dispersal of the seeds assists the plant in colonizing new areas. Plants advertise fruits by a ripening process in which the fruits change color, taste, and odor. Once fruits are ripe, they can be attacked by damaging agents, like fungi and bacteria, or by destructive feeders who will not disperse the seeds. How can plants encourage good seed predators who will disperse their seeds and at the same time discourage destructive seed predators?

There are four general ways for a plant to defend its fruit against damage by destructive herbivores. The cheapest defense is to fruit during the season of the year of minimum pest numbers. For example, in the temperate zone insect damage to fruits will be minimal in the autumn and winter. Second, plants can reduce their exposure to fruit damage by ripening fruit more quickly. This strategy, however, will also reduce the availability of ripe fruit to dispersers and may be counterproductive to the plant. Third, by making the fruit very unbalanced nutritionally, a plant may discourage insect and fungal attack. Fruit pulp is high in carbohydrates, low in lipids, and extremely low in protein. Fruits are among the poorest protein sources in nature, and this may be a general pest defense mechanism. But this strategy is somewhat dangerous because both good and bad seed predators are adversely affected by the low protein levels in the fruits.

Finally, plants may adopt a chemical defense of their fruits, although this is the most expensive method. Since chemical defenses in fruits reduce herbivore attack and thus seed dispersal, it has commonly been assumed that as fruits ripen, chemical defenses should be eliminated. For example, the alkaloid tomatine in green tomatoes is degraded by a new enzyme system that is activated in ripening red tomatoes. But the assumption that ripe fruits will not be chemically defended is probably a biased extrapolation from cultivated fruits that have been selected by humans for centuries. Of all the wild European plants that produce fleshy fruits, at least one-third have fruits toxic to humans. Toxic fruits are avoided by many birds and mammals, and consequently such fruits are dispersed only by a selected subset of vertebrates who can detoxify the specific chemicals. Toxic fruits are thus one way for a plant to "choose" its dispersal agents, although the detailed reasons why this is adaptive for the plants have not been discovered.

The interaction of seed predation and seed dispersal has been analyzed particularly clearly in the case of Clark's nutcracker (*Nucifraga columbiana*) and the whitebark pine *(Pinus albicaulis).* Several soft pines in North America have large, wingless seeds that are not dispersed by wind. These seeds are frequently removed from their cones by jays and nutcrackers, transported some distance, and buried in the soil as a future food source. When the birds cache more seeds than they can subsequently eat, the surplus is available for germination. This interaction can be considered a case of *mutualism* if both species profit from the association.

Does the whitebark pine benefit from seed dispersal from Clark's nutcracker? Nutcrackers cache three to five seeds, on average, in each store, bury them an average of 2 centimeters, avoid damp sites, and place many of their caches in microenvironments favorable for subsequent tree growth. One nutcracker stores about 32,000 whitebark pine seeds each year, which represents three to five times the energy required by the nutcracker. Thus many caches are not utilized, and the survival of pine seedlings arising from unused caches is high. Nutcrackers also disperse pine seeds to new areas within the subalpine forest zone, thus increasing the local distribution of whitebark pine. Many other species of birds feed on whitebark pine seeds, as do mice, chipmunks, and squirrels, but none of these species caches pine seeds in the proper way for germination. Thus the Clark's nutcracker–whitebark pine system seems to be a mutualism in which both species profit, the nutcracker by obtaining food and the pine by achieving seed dispersal.

Plants use a great variety of seed dispersal systems and much more work will have to be done before ecologists understand how many of these seed-eating interactions are mutualistic and how many are predatory.

Mutualism is perhaps the least studied of all the species interactions, and we do not know how significant mutualism is in determining community composition. A few examples have been worked out in detail and they may serve as a guide.

A cooperative system of defense has been evolved by the swollen-thorn acacias *(Acacia)* and their ant inhabitants *(Pseudomyrmex)* in the New World tropics. The ants depend on the acacia tree for food and a place to live, and the acacia depends on the ants for protection from herbivores and neighboring plants. Not all acacias (approximately 700 species) in the New World tropics depend on the ants, and not all the acacia ants (150 species or more) depend completely on acacia. In a few cases a high degree of mutualism has developed, described in detail by D. Janzen from his studies in Central America.

Swollen-thorn acacias have large, hollow thorns in which the ants live (Figure 10.10). The ants feed on modified leaflet tips called Beltian bodies, which are the primary source of protein and oil for the ants, and also on enlarged nectaries, which supply sugars. Swollen-thorn acacias maintain year-round leaf production, even in the dry season, to provide food for the ants.

The acacia ants continually patrol the leaves and branches of the acacia tree and immediately attack any herbivore that attempts to eat acacia leaves or bark. The ants also bite and sting any foreign vegetation that touches an acacia, and they clear all the vegetation from the ground beneath the acacia tree. Thus the swollen-thorn acacia often grows in a cylinder of space virtually free of all foreign vegetation. Some of the species of ants that inhabit acacia thorns live nowhere else.

If ants are removed from swollen-thorn acacias, the trees are quickly destroyed by herbivores or are crowded out by other plants. In his experimen-

(c)

(a)

(b)

Figure 10.10 Mutualism between acacias and their ant inhabitants. (a) *Acacia collinsii* growing in open pasture in Nicaragua. This tree had a colony of about 15,000 worker ants and was about 4 meters tall. (b) Area cleared over 10 years around a growing *Acacia collinsii* in Panama by ants chewing on all vegetation except the acacia. Machete in photo is 70 centimeters long. The area was not disturbed by other animals. (c) Swollen thorns of *Acacia cornigera* on a lateral branch. Each thorn is occupied by 20 to 40 immature ants and 10 to 15 worker ants. All the thorns on the tree are occupied by one colony. An ant entrance hole is visible in the left tip of the fourth thorn up from the bottom. (Photos courtesy of D. H. Janzen)

tal studies Janzen found that acacias without ants grew less and were often killed.

	Acacias with ants removed	**Acacias with ants present**
Survival rate over 10 months (%)	43	72
Growth increment (cm)		
May 25–June 16	6.2	33.0
June 16–August 3	10.2	72.9

Swollen-thorn acacias have apparently lost (or never had) the chemical defenses against herbivores found in other trees in the tropics. The ant-acacia system is a model system of the evolution of two species in an association of mutual benefit. The ants reduce herbivore destruction and competition from adjacent plants and thus serve as a living defense mechanism.

There is abundant evidence now that pairs of species or small groups of species may be tied together by evolutionary linkages through competition, predation, and mutualism. But how far does this extend? Are most species neutral with respect to each other so that community organization is loose? Or are the linkages stronger than we think so that community organization is tight?

One way to look at this difficult question is to see how convergent communities are. Are communities in similar climatic areas of different continents similar or has evolution produced different end points on different continents? If linkages are strong in natural communities, we would expect to find strong convergence in communities on different continents. Let us consider a few examples.

The Mediterranean type of climate, characterized by summer drought and winter rains, occurs in California, central Chile, South Africa, southern Australia, and around the Mediterranean Sea. In all of these areas the vegetation has a similar appearance—a dense scrub of woody, evergreen bushes with hard and stiff leaves. The plants of these areas are completely distinct, belonging to different taxonomic families, and no species occur in common among these areas. In counts of the total numbers of species, drought-avoiding annual plants predominate, and herbaceous perennial plants are also common (Figure 10.11).

The convergence of plant growth forms in communities on different continents was pointed out by plant geographers early in the nineteenth century, and there are similar convergences in animal form. But do other features of ecological communities converge as well? Some features do, but others do not.

Plant productivity is limited by temperature and rainfall in terrestrial communities all over the globe. We do not need to know on what continent

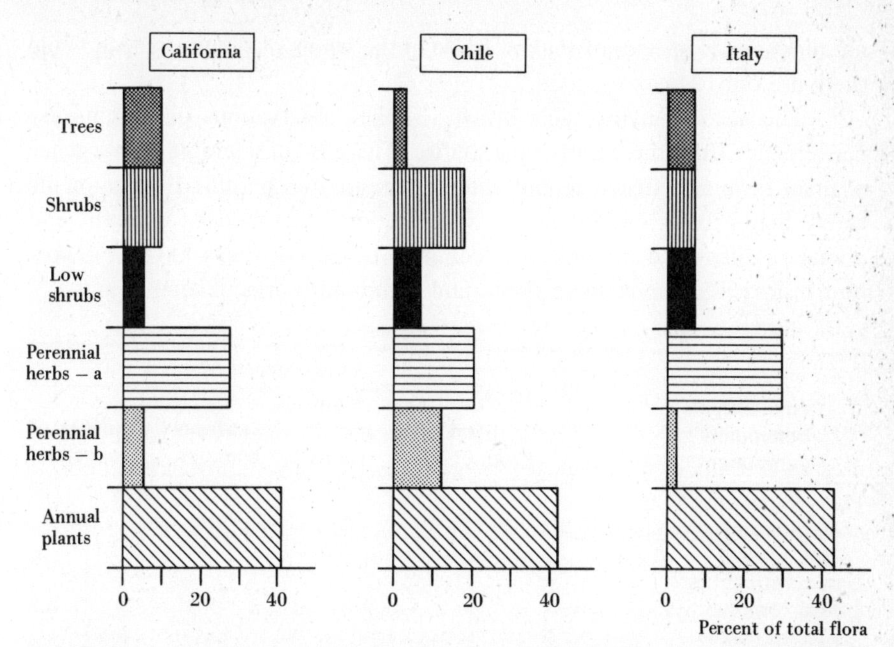

Figure 10.11 Convergence of plant growth forms in Mediterranean-type plant communities of California (391 species), Chile (319 species), and Italy (866 species). Drought-avoiding annuals predominate in all these communities. a = herbs with buds at the soil surface; b = herbs with buds under the soil surface. (Data from Mooney and Dunn 1970)

the community occurs or what plant species are involved to make a good prediction of plant productivity. The reason for this global generality is that photosynthesis is basic to all plants and is limited simply by physical factors—temperature and availability of water—in all areas.

Convergence in animal communities on different continents is present but less precise than that found in plants. If we compare Mediterranean communities of lizards and birds in California with their counterpart in Chile, we find these vertebrates are closely matched in most but not all habitats.

	Number of species	
	California	**Chile**
Lizards	5	4
Birds		
Total	69	39
Chaparral	17	19
Evergreen woodland	31	18

Small mammals, snakes, and insects, by contrast, are not well matched between Chile and California. For example, there are no seed-eating rodents in Chile that are like the pocket mice and kangaroo rats of California. In California there are 35 species of snakes, while in the whole of Mediterranean Chile there are only 2 snake species.

The accompanying table draws together observations on community convergence from these and other studies. There is considerable convergence of plant form and structure, but animal communities on different continents rarely match on a one-to-one basis. Evolution has produced different end points on different continents. The communities we see today have an important historical component to them, and community organization is loose.

Community or community component	Convergence of			
	Plant productivity	Plant growth forms	Seasonal changes	Number of species
Plant communities				
Mediterranean	Yes	Yes	Yes	No
Hot desert	—	Yes	—	—
Bird communities				Yes
Lizard communities				
Mediterranean				Yes
Hot desert				No
Seed-eating animals of hot deserts				No
Leaf-eating insects				No
Flower visitors in hot deserts				No

Source: From Orians and Paine, 1983.

An important conservation message can be derived from this analysis of community convergence: *Communities are unique and are not replaceable.* We cannot move animals and plants from one continent to another the same way we can swap computer chips or wood screws. They are simply not interchangeable, and so we should preserve what we have.

We conserve communities by setting aside national parks, but it is important to remember that our existing parks are legally defined and not ecologically defined. One important question for conservationists is to ask how big our parks ought to be to maintain the existing ecological community. The size of national parks must be defined in relation to the largest animals inhabiting the park, and the concern is that a minimum population size must be maintained to prevent genetic deterioration through inbreeding. The best estimates are that 50 individuals are needed for short-term survival of a species and 500 individuals for long-term ($\gg$ 1000 years) survival. If we accept these estimates, we can estimate the size of national parks required to hold this many individuals.

Table 10.1 on page 166 gives estimates of the ecological area required for eight national parks from western North America. Only two of these parks are large enough to support a minimal population of large mammals for short-term survival, and all of these parks should be much larger than they are to support populations sufficient for long-term survival. Figure 10.12 illustrates this discrepancy between legal and ecological boundaries for Yellowstone-Grand Teton National Park. Our national parks, which once seemed so large when they were first established, are clearly too small to maintain the existing ecological communities. The important message is that the management of large animals in existing parks must be done in cooperation with the surrounding land owners.

Do studies of evolutionary ecology give us any messages for designing our agricultural or forestry systems? The most important quality that humans should desire in agricultural systems is *stability.* What produces stability in

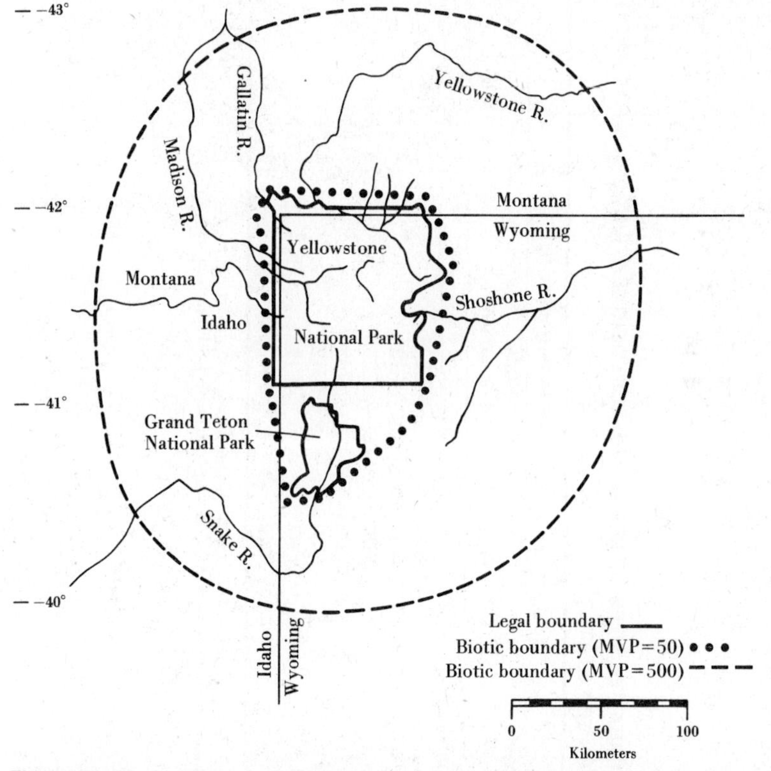

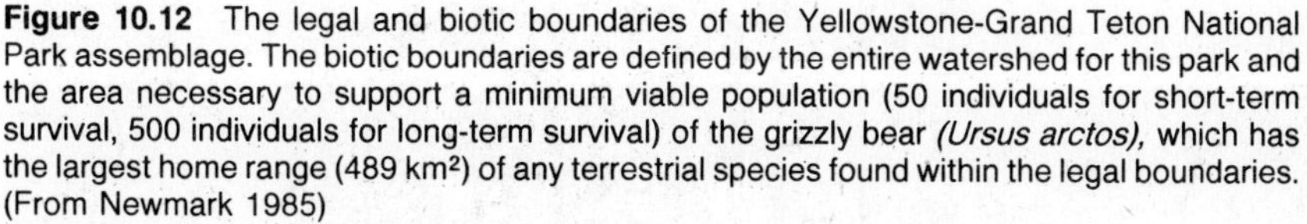

Figure 10.12 The legal and biotic boundaries of the Yellowstone-Grand Teton National Park assemblage. The biotic boundaries are defined by the entire watershed for this park and the area necessary to support a minimum viable population (50 individuals for short-term survival, 500 individuals for long-term survival) of the grizzly bear *(Ursus arctos),* which has the largest home range (489 km²) of any terrestrial species found within the legal boundaries. (From Newmark 1985)

TABLE 10.1 COMPARATIVE SIZES OF LEGAL AND ECOLOGICAL BOUNDARIES FOR NATIONAL PARKS IN WESTERN CANADA AND THE UNITED STATES

National park/park assemblage	Size of boundary (km^2)			Ratio	Species used to define biotic boundary
	Legal	Biotic (MVP = 50)	Biotic (MVP = 500)		
Kootenay-Banff-Jasper-Yoho	20736	12233	122330	1:0·6:6	Grizzly bear *Ursus arctos*
Grand Teton-Yellowstone	10328	12233	122330	1:1·2:12	Grizzly bear *Ursus arctos*
Grand Canyon	4931	8125	81250	1:1·6:16	Mountain lion *Felix concolor*
Glacier-Waterton Lakes	4627	12233	122330	1:2·6:26	Grizzly bear *Ursus arctos*
Olympic	3628	7490	74900	1:2·1:21	Mountain lion *Felix concolor*
Sequoia-Kings Canyon	3389	10125	101250	1:3·0:30	Wolverine *Gulo luscus*
Yosemite	2083	10125	101250	1:4·8:48	Wolverine *Gulo luscus*
Rocky Mountain	1049	10125	101250	1:9·6:96	Wolverine *Gulo luscus*

MVP = minimum viable population. The smaller biotic boundary will maintain a population of 50 individuals of the large mammals, enough for short-term survival. The larger biotic boundaries will maintain 500 individuals needed for long-term survival. (From Newmark 1985.)

natural communities? Stability both in crop systems and in natural communities means restricted fluctuations in pest populations over time. Instability means times of heavy pest damage in crops. Crops are much less stable systems than natural ecosystems. We worry much more about the pests of corn plants in Illinois farmland than we do about the pests of the adjacent deciduous forest communities. Why should this be?

Natural systems like forests are more stable than crop systems because their component species have had a long shared evolutionary history. Natural selection has eliminated plants that do not have adequate defenses against herbivores and predators that are too inefficient at locating prey. Crops by contrast have undergone selective breeding and have usually lost their genetic defense mechanisms. Human interference in the form of pesticides and fertilizers is constantly changing so that no coevolution has time to occur in crops.

This fundamental difference in evolutionary history between natural communities and crop systems means that we must rely on additional techniques to prevent pest outbreaks in agriculture. Physical complexity of the environment is the best barrier to pests, and the best advice is to grow crops in patches rather than in large blocks of single species. Patchiness or diversity in agriculture can be of two types. Interplanting crops is the most obvious type. In the Central Valley of California farmers can minimize pest problems with cotton by planting alfalfa strips along cotton fields. Lygus bugs are attracted out of cotton, where they do serious damage, into alfalfa where they do little harm. Chinese farmers have developed strip and small patch planting of many crops and have put these ecological ideas into practice much more than have Western farmers.

A second type of diversity is genetic diversity. In natural communities individuals of the same species are genetically variable. But crops are often genetically homogeneous, and pests can run wild. One important means of reducing crop pests would be to deliberately increase the number of genetic varieties used. Unfortunately, most of modern agriculture has been moving in the opposite direction.

Evolution is the net result of ecology in action. In the 130 years since Charles Darwin published *The Origin of Species* biologists have discovered how much natural selection has molded organisms to adapt them to the local physical and biological environment. There is no reason to think that we have discovered all the complex variations that selection has produced, particularly in tropical communities. We should be humbled by all this, and we should in the end come to the true message of ecology. *We are one species in a world whose biological heritage is as irreplaceable as our own cultural heritage. We are the custodians of this biological heritage and our goals must be to study it, to understand it, to enjoy it, and to pass it to our children undiminished.*

FURTHER READING

Boucher, D. H. (ed.). 1985. *The Biology of Mutualism; Ecology and Evolution.* Croom Helm, London.

*Dawkins, R. 1976. *The Selfish Gene.* Oxford University Press, New York.

Denno, R. F., and M. S. McClure (eds.). 1983. *Variable Plants and Herbivores in Natural and Managed Systems.* Academic Press, New York.

Endler, J. A. 1986. *Natural Selection in the Wild.* Princeton University Press, Princeton, N.J.

Frankel, O. H., and M. E. Soule. 1981. *Conservation and Evolution.* Cambridge University Press, London.

Futuyma, D. J., and M. Slatkin (eds.). 1983. *Coevolution.* Sinauer, Sunderland, Mass.

Shorrocks, B. (ed.). 1984. *Evolutionary Ecology.* Blackwell, Oxford.

Soule, M. E. (ed.) 1986. *Conservation Biology.* Sinauer, Sunderland, Mass.

*Highly recommended.

Glossary

acid rain Natural precipitation with pH less than 5.6 due to the presence of varying amounts of nitric and sulfuric acids resulting from the burning of oil and coal.

aggregation Coming together of organisms into a group, as in locusts.

algae A large group of relatively simple plants (lacking leaves, stems, and roots), mostly aquatic, with simply organized reproductive organs.

alkaloid Natural organic base found in plants; may be toxic to animals.

anaerobic Without oxygen.

annual Plant or animal that completes its life cycle in one year.

association Major unit in community ecology, characterized by essential uniformity of species composition.

biogeography Branch of biology that deals with the geographical distribution of plants and animals.

biological control Use of organisms or viruses to control parasites, weeds, or other pests.

biosphere The whole earth ecosystem.

biota Species of all plants and animals occurring within a certain area or region.

blue-green algae A group of algae, unicellular or filamentous, usually bluish-green in color.

bog A quagmire or wet, spongy ground where any heavy body is likely to sink; usually filled with peat.

boreal forest The spruce forests of northern United States, Canada, the Soviet Union, and Fennoscandia. Also called *taiga.*

Carboniferous period The geological period during which much coal was deposited, approximately 250 million years ago.

carnivore Flesh eater; organism that eats animals.

cellulose A carbohydrate forming the main part of plant cell walls; difficult for many animals to digest.

climax Kind of community capable of perpetuation under the prevailing climatic and soil conditions.

coevolution Reciprocal evolutionary influences between two or more interacting species, e.g., between predator and prey.

community Group of populations of plants and animals in a given place; ecological unit used in a broad sense to include groups of various sizes and degrees of integration.

competition Common use of a resource by a number of organisms of the same or of different species when the resource is in short supply (exploitation); common use of a resource, regardless of supply when the organisms seeking that resource harm one another in the process (interference).

conifer A large group, about 400 species, of woody and mostly evergreen trees and shrubs, with the reproductive organs in cones and often relatively soft wood.

cursorial Adapted for running.

deciduous Falling at the end of the growth period, as the leaves of many temperate zone trees.

density Number of individuals in relation to the space in which they occur.

detritus Fragments of dead organisms, either fresh or partly decomposed.

diatom A unicellular form of algae with cell walls impregnated with silica.

dispersal The movement or scattering of individuals away from their place of birth or residence.

dominance Condition in communities or in vegetational strata in which one or more species, by means of their number, coverage, or size, have considerable influence upon or control of the conditions of existence of associated species.

ecology The scientific study of the interactions that determine the distribution and abundance of organisms.

ecosystem Biotic community and its abiotic environment.

ecotone Transition zone between two diverse communities (e.g., the tundra–boreal forest ecotone).

ecotype A habitat type; a race of a species adapted to a specific kind of local environment.

edaphic Pertaining to the soil.

El Niño A change in the ocean upwelling pattern off Peru and Ecuador during which warm water invades and replaces the normal cold water.

emigration The movement of organisms out of a habitat or defined area.

environment All the biotic and abiotic factors that actually affect an individual organism at any point in its life cycle.

epidemiology Branch of medicine dealing with the incidence and distribution of disease in a population.

eutrophic Referring to a substrate for plant growth, either a body of water or soil, that is nutrient-rich and highly productive.

eutrophication The process of nutrient enrichment of lakes by human activities, often leading to habitat degradation.

exotic species A species that is not native to a particular geographical area, usually from another continent.

fecundity The number of offspring produced during a unit of time.

fertility A physiological notion indicating that an organism is capable of breeding.

food chain (or food web) Figure of speech for the dependence for food of organisms upon others in a series, beginning with plants and ending with the largest carnivores.

fry Newly hatched young of fishes.

fugitive species Species of plants or animals with high dispersal abilities that colonize vacant areas rapidly.

gall An abnormal growth on a plant caused by the attack of a parasite.

genotype Characters of an organism due to its genetic constitution; inherited characteristics.

global stability Ability to return to an original configuration after a perturbation of large magnitude. Compare *local stability.*

Gondwanaland The great southern continent that broke apart about 200 million years ago to form the modern continents.

greenhouse effect Any physical process that allows heat to come into an area but prevents some of it from leaving, thereby warming the area.

gregarious Tending to herd together or to form a group.

habitat An environment which a species or a whole community occupies.

hardwood A large group of woody, mostly large, and usually deciduous trees, with relatively hard wood.

herbaceous Having little or no woody stem.

herbivore Organism that eats plants.

homeostasis Maintenance of a high degree of uniformity in functions of an organism or interactions of individuals in a population or community under changing conditions, because of the capabilities of organisms to make adjustments.

host Organism that furnishes food, shelter, or other benefits to another organism of a different species.

humus Material formed by the decomposition of plant or animal matter, constituting the organic, nutrient-rich, part of the soil.

immigration The movement of individuals into a defined habitat or area.

indicator species A form of organism or a species indicating the presence of certain environmental conditions.

interspecific competition Competition between members of different species.

intertidal The zone along the sea shore between the low and high tide lines.

invasion area In locusts, an area colonized by swarms during an outbreak, but incapable of supporting permanent populations.

keystone species Species in a community whose removal provokes a catastrophic change in the composition and structure of the rest of the community.

larva An immature stage of insect development between the egg and the adult.

leaching The removal, by percolating water, of mineral salts from the soil.

lignin A complex substance which, together with cellulose, causes the thickening of plant cell walls and so forms wood.

litter The upper, slightly decomposed, layer of the forest floor.

littoral Shallow-water zone of lakes or the sea, with light penetration to the bottom; often occupied by rooted aquatic plants.

local stability Ability to return to an original configuration after a perturbation of small magnitude. Compare *global stability.*

migration The movement of organisms from one habitat or region to another.

mimicry The assumption of the habits, color, or structure of another organism, usually for protection.

monoculture An extensive area dominated by a single species of plant or animal.

moor Open or treeless area dominated by shrubs and grasses.

moraine A trail of rock waste left piled at the margins of a glacier.

muskeg Moss bogs of the spruce forests of Canada and the Soviet Union.

mutualism Interaction between two species in which both benefit from the association.

myxomatosis A virus disease native to South American rabbits, introduced to Australia and Europe to control the European rabbit.

nannoplankton Microscopic floating plants and animals, among the smallest plankton, about 2 to 20 microns long.

net production Production after respiration losses are subtracted. See *production.*

niche Role of an organism in the environment; the activities and relationships of an organism in the community.

nomadic Leading a wandering life; moving in an irregular pattern.

obligate limited to one mode of life or action; not optional.

obligate predator or **parasite** Predator or parasite that is restricted to a single species of prey or host.

oligotrophic Referring to a substrate for plant growth, either a body of water or soil, that is nutrient-poor and not productive.

optimum yield Amount of material that can be removed from a population that will maximize biomass (or numbers or profit or any other type of "optimum") on a sustained basis.

outbreak area In locusts, an area that can produce swarms that may then move into a temporary invasion area.

ozone (O_3) A powerful oxidizing agent, produced when oxygen is exposed to an electrical discharge.

parasite Organism that feeds upon, secures shelter from, or derives other benefits from association with another organism (the host) and injures the host in the process.

pelagic of or living in the open sea; referring to birds, fishes, plankton, and other organisms that live in surface water and do not depend on the bottom.

perennial Plant that persists for a number of years.

permafrost Permanently frozen layer of soil; above the permafrost a thin layer of soil may freeze and thaw each year.

pH A scale for expressing acidity or alkalinity of a solution with pH 7 neutral, above 7 more alkaline, and below 7 more acidic.

phenotype Characters of an organism that are due to environmental variations acting on genetic constitution during development.

photosynthesis Synthesis of carbohydrates from carbon dioxide and water by chlorophyll using light as energy with oxygen as a by-product.

phytoplankton Plant portion of the plankton; contains many species of algae and diatoms. See *plankton.*

pioneer species Species that colonize new areas quickly and are thus typically the first organisms in a succession; usually species with small seeds or propagules.

plague An outbreak or sudden superabundance of a pest.

plankton Small, floating or weakly swimming plants and animals in freshwater and marine situations.

population Group of individuals of a single species.

primary production Production by green plants using the process of photosynthesis. See *production.*

production Amount of energy (or material) formed by an individual, population, or community in a specific time period; includes growth and reproduction only.

protozoan A group of simple, usually very small unicellular organisms.

recruitment Increment to a natural population by birth or migration, usually from young animals or plants entering the adult population.

respiration Complex series of chemical reactions in all organisms in which carbon dioxide, water, and energy are end products.

saprophyte Plant that obtains food from dead or decaying organic matter.

savannah An open grassland-woodland habitat in tropical or subtropical regions.

secondary plant substance Chemical found in plants that has no known metabolic function and is often toxic to other organisms.

secondary production Production by consumer organisms—herbivores, carnivores, or detritus feeders. See *production.*

spawn Product or offspring; eggs deposited by aquatic animals.

stability Absence of fluctuations in populations; ability to withstand perturbations without large changes in composition.

steppe Extensive area of natural, dry grassland; usually used in reference to grasslands in southwestern Asia and southeastern Europe and equivalent to prairie in North American usage.

subtidal The zone along the sea shore below the lowest tide line.

succession Replacement of one kind of community by another kind; the progressive changes in vegetation and animal life that may culminate in the stable climax community.

symbiosis In a broad sense, the living together of two or more organisms of different species; in a narrow sense, synonymous with mutualism.

taiga The northern boreal forest zone, a broad band of coniferous forest south of the arctic tundra.

tannin A chemical substance found in plants that reduces digestibility.

terpene A group of essential oils found in plants.

terrestrial Of dry ground.

territory An area defended by an animal, often for breeding or feeding.

till A soil of mixed clay and pebbles, derived from glaciation.

trace element Chemical element used by organisms in minute quantities and essential to their physiology.

trophic level Functional classification of organisms in a community according to feeding relationships, from first trophic level of green plants through succeeding levels of herbivores and carnivores.

tundra Treeless area in arctic and alpine regions, varying from bare ground to various types of vegetation consisting of grasses, sedges, forbs, dwarf shrubs, lichens, and mosses.

ungulate An order of terrestrial herbivorous mammals with hoofs (e.g., goats, deer, horses, antelope).

vector Organism (often an insect) that transmits a pathogen from one organism to another.

zooplankton Animal portion of the plankton. See *plankton.*

References

INTRODUCTION

Labisky, R. F. 1975. Illinois pheasants: their distribution and abundance, 1958–1973. *Illinois Natural History Survey Biological Notes* No. 94:1–11.

Warner, R. E. 1981. Illinois pheasants: population, ecology, distribution, and abundance, 1900–1978. *Illinois Natural History Survey Biological Notes* No. 115: 1–22.

CHAPTER 1

Borchert, J. R. 1950. The climate of the central North American grassland. *Annals of the Association of American Geographers* 40:1–39.

Daubenmire, R. F. 1954. Alpine timberlines in the Americas and their interpretation. *Butler University Botanical Studies* 2:119–136.

Johnson, S. R., and I. McT. Cowan. 1974. Thermal adaptation as a factor affecting colonizing success of introduced Sturnidae (Aves) in North America. *Canadian Journal of Zoology* 52:1559–1576.

Kessel, B. 1953. Distribution and migration of the European starling in North America. *Condor* 55:49–67.

Kitching, J. A., and F. J. Ebling. 1967. Ecological studies at Lough Ine. *Advances in Ecological Research* 4:197–291.

MacCrimmon, H. R. 1971. World distribution of rainbow trout *(Salmo gairdneri)*. *Journal of the Fisheries Research Board of Canada* 28:663–704.

Myers, K. 1971. The rabbit in Australia. In *Dynamics of Populations,* ed. P. J. den Boer and G. R. Gradwell, pp. 478–506. Proceedings of the Advanced Study Institute on the Dynamics of Numbers in Populations (Oosterbeek), Wageningen.

Tranquillini, W. 1979. *Physiological Ecology of the Alpine Timberline.* Springer-Verlag, Berlin.

Weaver, J. E. 1968. *Prairie Plants and Their Environment.* University of Nebraska Press, Lincoln.

CHAPTER 2

Ebling, F. J., and D. M. Stoddart (eds.). 1977. *Population Control by Social Behaviour.* Institute of Biology, London.

Jenkins, D., and A. Watson. 1970. Population control in red grouse and rock ptarmigan in Scotland. *Transactions of the Congress of the International Union of Game Biologists* 7:121–141.

Jenkins, D., A. Watson, and G. R. Miller. 1967. Population fluctuations in the red grouse, *Lagopus lagopus scoticus. Journal of Animal Ecology* 36:97–122.

Krebs, C. J. 1964. The lemming cycle at Baker Lake, Northwest Territories, during 1959–62. *Arctic Institute of North America,* Technical Paper No. 15:1–104.

Krebs, C. J., and J. H. Myers. 1974. Population cycles in small mammals. *Advances in Ecological Research* 8:268–399.

Miller, G. R., and A. Watson. 1974. Heather moorland: a man made ecosystem. In *Conservation in Practice,* ed. A. Warren and F. B. Goldsmith, pp. 145–166. Wiley, London.

Moss, R., A. Watson, and P. Rothery. 1984. Inherent changes in the body size, viability and behaviour of a fluctuating red grouse *(Lagopus lagopus scoticus)* population. *Journal of Animal Ecology* 53:171–189.

Murton, R. K. 1965. *The Wood-pigeon.* Collins, London.

Murton, R. K. 1965. Natural and artificial population control in the wood pigeon. *Annals of Applied Biology* 55:177–192.

Murton, R. K. 1971. *Man and Birds.* Collins, London.

Rainey, R. C., E. Betts, and A. Lumley. 1979. The decline of the desert locust plague in the 1960s: control operations or natural causes? *Philosophical Transactions of the Royal Society of London,* Series B, 287:315–344.

Schultz, A. M. 1969. A study of an ecosystem: the arctic tundra. In *The Ecosystem Concept in Natural Resource Management,* ed. G. van Dyne, pp. 77–93. Academic Press, New York.

Sheail, J. 1971. *Rabbits and Their History.* David and Charles, Newton Abbot.

Uvarov, B. P. 1977. *Grasshoppers and Locusts. A Handbook of General Acridology.* Vol. 2. Cambridge University Press, London.

Waloff, Z. 1976. Some temporal characteristics of Desert Locust plagues. *Anti-Locust Memoir* 13:1–36.

Watson, A., and G. R. Miller. 1971. Territory size and aggression in a fluctuating red grouse population. *Journal of Animal Ecology* 40:367–383.

CHAPTER 3

Campbell, H., D. K. Martin, P. E. Fukovich, and B. K. Harris. 1973. Effect of hunting and some other environmental factors on scaled quail in New Mexico. *Wildlife Monographs* 34:1–49.

Caughley, G., G. C. Grigg, J. Caughley, and G. J. E. Hill. 1980. Does dingo predation control the densities of kangaroos and emus? *Australian Wildlife Research* 7:1–12.

Debach, P. 1974. *Biological Control by Natural Enemies.* Cambridge University Press, London.

Feeny, P. P. 1970. Seasonal changes in oak leaf tannins and nutrients as a cause of spring feeding by winter moth caterpillars. *Ecology* 51:565–581.

Hartesveldt, R. J., and H. T. Harvey. 1967. The fire ecology of sequoia regeneration. *Proceedings of the Tall Timbers Fire Ecology Conference* 7:65–77.

Hewson, R. 1976. Grazing by mountain hares *Lepus timidus* L., red deer *Cervus elaphus* L. and red grouse *Lagopus l. scoticus* on heather moorland in northeast Scotland. *Journal of Applied Ecology* 13:657–666.

Jenkins, D., A. Watson, and G. R. Miller. 1970. Practical results of research for management of red grouse. *Biological Conservation* 2:266–272.

Kalleberg, H. 1958. Observations in a stream tank of territoriality and competition in juvenile salmon and trout (*Salmo salar* L. and *S. trutta* L.). *Institute of Freshwater Research,* Drottningholm, Report No. 39:55–98.

Kilgore, B. M. 1972. Impact of prescribed burning on a sequoia–mixed conifer forest. *Proceedings of the Tall Timbers Fire Ecology Conference* 12:345–375.

Klein, D. R. 1965. Ecology of deer range in Alaska. *Ecological Monographs* 35:259–284.

Murdoch, W. W. 1975. Diversity, complexity, stability, and pest control. *Journal of Applied Ecology* 12:795–807.

Newsome, A. E. 1975. An ecological comparison of the two arid-zone kangaroos of Australia, and their anomalous prosperity since the introduction of ruminant stock to their environment. *Quarterly Review of Biology* 50:389–424.

Painter, R. H. 1951. *Insect Resistance in Crop Plants.* Macmillan, New York.

Redfield, J. A., F. C. Zwickel, and J. F. Bendell. 1970. Effects of fire on numbers of blue grouse. *Proceedings of the Tall Timbers Fire Ecology Conference* 10:63–83.

Rosenthal, G. A., and D. H. Janzen (eds.). 1979. *Herbivores: Their Interaction with Secondary Plant Metabolites.* Academic Press, New York.

Snyder, W. D. 1967. Experimental habitat improvement for scaled quail. *Colorado Game, Fish and Parks Department Technical Publication* No. 19:1–65.

Tatum, L. A. 1971. The Southern corn leaf blight epidemic. *Science* 171:1113–1116.

van den Bosch, R. 1976. Three generations of pesticides: time for a new approach. *Transactions of the 41st North American Wildlife and Natural Resources Conference,* pp. 51–57.

von Haartman, L. 1971. Population dynamics. In *Avian Biology,* vol. 1, ed. D. S. Farner and J. R. King. Academic Press, New York.

Watson, A., R. Hewson, D. Jenkins, and R. Parr. 1973. Population densities of mountain hares compared with red grouse on Scottish moors. *Oikos* 24:225–230.

CHAPTER 4

Agenbroad, L. D. 1984. New World mammoth distribution. In *Quaternary Extinctions,* ed. P. S. Martin and R. G. Klein, pp. 90–108. University of Arizona Press, Tucson.

Allen, K. R. 1980. *Conservation and Management of Whales.* University of Washington Press, Seattle.

Beddington, J. R., and R. M. May. 1982. The harvesting of interacting species in a natural ecosystem. *Scientific American* 247(5):42–49.

Burch, E. S., Jr. 1977. Muskox and man in the central Canadian subarctic, 1689–1974. *Arctic* 30:135–154.

Cooley, R. A. 1963. *Politics and Conservation: The Decline of the Alaska Salmon.* Harper & Row, New York.

Duffy, D. C. 1983. Environmental uncertainty and commercial fishing: effects on Peruvian guano birds. *Biological Conservation* 26:227–238.

Fisher, J. and R. M. Lockley. 1954. *Sea-birds.* Collins, London.

Fisher, J., and H. G. Vevers. 1944. The breeding distribution, history and population of the North Atlantic gannet *(Sula bassana). Journal of Animal Ecology* 13:49–62.

Gordon, B. C. 1977. Muskox and man in the subarctic: an archaeological view. *Arctic* 30:246.

Hardin, G. 1968. The tragedy of the commons. *Science* 162:1243–1246.

Hester, J. J. 1967. The agency of man in animal extinctions. In *Pleistocene Extinctions,* ed. P. S. Martin and H. E. Wright, Jr., pp. 169–192. Yale University Press, New Haven, Conn.

LeBrasseur, R. J., C. D. McAllister, and T. R. Parsons. 1979. Addition of nutrients to a lake leads to greatly increased catch of salmon. *Environmental Conservation* 6:187–190.

Martin, P. S. 1967. Prehistoric overkill. In *Pleistocene Extinctions,* ed. P. S. Martin and H. E. Wright, Jr., pp. 75–120. Yale University Press, New Haven, Conn.

Martin, P. S. 1984. Prehistoric overkill: the global model. In *Quaternary Extinctions,* ed. P. S. Martin and R. G. Klein, pp. 354–403. University of Arizona Press, Tucson.

Martin, P. S., and J. E. Guilday. 1967. A bestiary for Pleistocene biologists. In *Pleistocene Extinctions,* ed. P. S. Martin and H. E. Wright, Jr., pp. 1–62. Yale University Press, New Haven, Conn.

Martin, P. S. and R. G. Klein (eds.). 1984. *Quaternary Extinctions: A Prehistoric Revolution.* University of Arizona Press, Tucson.

Martin, P. S., and H. E. Wright, Jr. (eds.). 1967. *Pleistocene Extinctions. The Search for a Cause.* Yale University Press, New Haven, Conn.

Nelson, J. B. 1978. *The Sulidae: Gannets and Boobies.* Oxford University Press, London.

Ricker, W. E. 1973. Two mechanisms that make it impossible to maintain peak-period yields from stocks of Pacific Salmon and other fishes. *Journal of the Fisheries Research Board of Canada* 30:1275–1286.

Ritchie, J. 1920. *The Influence of Man on Animal Life in Scotland.* Cambridge University Press, Cambridge.

Russell, E. S. 1931. Some theoretical considerations on the "overfishing" problem. *Journal of Cons. Perm. Int. Exp. Mer* 6:3–27.

Schaefer, M. B. 1970. Men, birds and anchovies in the Peru Current—dynamic interactions. *Transactions of the American Fisheries Society* 99:461–467.

Silliman, R. P., and J. S. Gutsell. 1958. Experimental exploitation of fish populations. *Fishery Bulletin* 58(133):215–252.

CHAPTER 5

Barraclough, W. E., and D. Robinson. 1972. The fertilization of Great Central Lake. III. Effect on juvenile sockeye salmon. *Fishery Bulletin* 70:37–48.

Billings, W. D. 1938. The structure and development of old-field shortleaf pine stands and certain associated physical properties of the soil. *Ecological Monographs* 8:437–499.

Bormann, F. H., and G. E. Likens. 1979. Catastrophic disturbance and the steady state in northern hardwood forests. *American Scientist* 67:660–669.

Burk, C. J. 1977. A four year analysis of vegetation following an oil spill in a freshwater marsh. *Journal of Applied Ecology* 14:515–522.

Connell, J. H., and R. O. Slatyer. 1977. Mechanisms of succession in natural communities and their role in community stability and organization. *American Naturalist* 111:1119–1144.

Connell, J. H., and W. P. Sousa. 1983. On the evidence needed to judge ecological stability or persistence. *American Naturalist* 121:789–824.

Crocker, R. L., and J. Major. 1955. Soil developments in relation to vegetation and surface age at Glacier Bay, Alaska. *Journal of Ecology* 43:427–448.

Dunnet, G. M. 1982. Oil pollution and seabird populations. *Philosophical Transactions of the Royal Society of London,* Series B, 297:413–427.

Edmondson, W. T., and J. T. Lehman. 1981. The effects of changes in the nutrient income on the condition of Lake Washington. *Limnology and Oceanography* 26:1–29.

Edmondson, W. T., and A. H. Litt. 1982. *Daphnia* in Lake Washington. *Limnology and Oceanography* 27:272–293.

Holdgate, M. W. 1976. Closing summary. In *Marine Ecology and Oil Pollution,* ed. J. M. Baker. Applied Science Publications, Barking, Essex, England.

Hyatt, K. D., and J. G. Stockner. 1985. Responses of sockeye salmon *(Oncorhynchus nerka)* to fertilization of British Columbia coastal lakes. *Canadian Journal of Fisheries and Aquatic Sciences* 42:320–331.

Keever, C. 1950. Causes of succession on old fields of the Piedmont, North Carolina. *Ecological Monographs* 20:229–250.

Kilgore, B. M. 1984. Restoring fire's natural role in America's wilderness. *Western Wildlands* 11(3):2–8.

Lawrence, D. B. 1958. Glaciers and vegetation in southeastern Alaska. *American Scientist* 46:89–122.

McIntyre, A. D. 1982. Oil pollution and fisheries. *Philosophical Transactions of the Royal Society of London,* Series B, 297:401–411.

Southward, A. J. 1982. An ecologist's view of the implications of the observed physiological and biochemical effects of petroleum compounds on marine organisms and ecosystems. *Philosophical Transactions of the Royal Society of London,* Series B, 297:241–255.

Stone, E. C., and R. B. Vasey. 1968. Preservation of coast redwood on alluvial flats. *Science* 159:157–161.

Tilman, D. 1985. The resource-ratio hypothesis of plant succession. *American Naturalist* 125:827–852.

Williamson, G. B., and E. M. Black. 1981. High temperature of forest fires under pines as a selective advantage over oaks. *Nature* 293:643–644.

CHAPTER 6

Aron, J. L., and R. M. May. 1982. The population dynamics of malaria. In *The Population Dynamics of Infectious Diseases: Theory and Applications,* ed. R. M. Anderson. Chapman and Hall, London.

Briand, F. 1983. Environmental control of food web structure. *Ecology* 64:253–263.

Brooks, J. L., and S. I. Dodson. 1965. Predation, body size, and composition of plankton. *Science* 150:28–35. Copyright 1965 by the American Association for the Advancement of Science.

Busvine, J. R. 1975. *Arthropod Vectors of Disease.* Edward Arnold, London.

Everson, I. 1977. *The Living Resources of the Southern Ocean.* UN Development Program, Food and Agriculture Organization of the United Nations, Rome.

Laws, R. M. 1985. The ecology of the Southern Ocean. *American Scientist* 73:26–40.

May, R. M. 1977. Thresholds and breakpoints in ecosystems with a multiplicity of stable states. *Nature* 269:471–477.

Parsons, T. R., and M. Takahashi. 1973. *Biological Oceanographic Processes.* Pergamon Press, Oxford.

Pitelka, F. A., P. Q. Tomich, and G. W. Treichel. 1955. Ecological relations of jaegers and owls as lemming predators near Barrow, Alaska. *Ecological Monographs* 25:85–117.

Regier, H. A. 1973. The sequence of exploitation of stocks in multi-species fisheries in the Laurentian Great Lakes. *Journal of the Fisheries Research Board of Canada* 30:1992–1999.

Regier, H. A., and W. L. Hartman. 1973. Lake Erie's fish community: 150 years of cultural stresses. *Science* 180:1248–1255. Copyright 1973 by the American Association for the Advancement of Science.

Schindler, D. W. 1974. Eutrophication and recovery in experimental lakes: implications for lake management. *Science* 184:897–899. Copyright 1974 by the American Association for the Advancement of Science.

Schindler, D. W. 1978. Factors regulating phytoplankton production and standing crop in the world's freshwaters. *Limnology and Oceanography* 23:478–486.

Smith, V. H. 1983. Low nitrogen to phosphorus ratios favor dominance by blue-green algae in lake phytoplankton. *Science* 221:669–671.

Thomas, A. S. 1960. Changes in vegetation since the advent of myxomatosis. *Journal of Ecology* 48:287–306.

Watson, A. 1983. Eighteenth century deer numbers and pine regeneration near Braemar, Scotland. *Biological Conservation* 25:289–305.

Watt, A. S. 1962. The effect of excluding rabbits from grassland A (Xerobrometum) in Breckland, 1936–60. *Journal of Ecology* 50:181–198.

Woodwell, G. M. 1967. Toxic substances and ecological cycles. *Scientific American* 216(3):24–31.

CHAPTER 7

Caughley, G. 1976. The elephant problem—an alternative hypothesis. *East Africa Wildlife Journal* 14:265–283.

Dayton, P. K., V. Currie, T. Gerrodette, B. D. Keller, R. Rosenthal, and D. Ven Tresca. 1984. Patch dynamics and stability of some California kelp communities. *Ecological Monographs* 54:253–289.

Duggins, D. O. 1983. Starfish predation and the creation of mosaic patterns in a kelp-dominated community. *Ecology* 64:1610–1619.

Eltringham, S. K. 1982. *Elephants.* Blandford Press, Poole, Dorset.

Estes, J. A., R. J. Jameson, and E. B. Rhode. 1982. Activity and prey selection in the sea otter: influence of population status on community structure. *American Naturalist* 120:242–258. Copyright 1982 by the University of Chicago.

Hatton, J. C., and N. O. E. Smart. 1984. The effect of long-term exclusion of large herbivores on soil nutrient status in Murchison Falls National Park, Uganda. *African Journal of Ecology* 22:23–30.

Laws, R. M. 1970. Elephants as agents of habitat and landscape change in East Africa. *Oikos* 21:1–15.

Mann, K. H. 1985. Invertebrate behaviour and the structure of marine benthic communities. In *Behavioural Ecology,* ed. R. M. Sibley and R. H. Smith. Blackwell, Oxford.

Mann, K. H., and P. A. Breen. 1972. The relation between lobster abundance, sea urchins, and kelp beds. *Journal of the Fisheries Research Board of Canada* 29:603–605.

Paine, R. T. 1966. Food web complexity and species diversity. *American Naturalist* 100:65–75.

Paine, R. T. 1974. Intertidal community structure. Experimental studies on the relationship between a dominant competitor and its principal predator. *Oecologia* 15:93–120.

Paine, R. T., J. C. Castillo, and J. Cancino. 1985. Perturbation and recovery patterns of starfish-dominated intertidal assemblages in Chile, New Zealand, and Washington State. *American Naturalist* 125:679–691.

Paine, R. T., and S. A. Levin. 1981. Intertidal landscapes: disturbance and the dynamics of pattern. *Ecological Monographs* 51:145–178.

Pimm, S. L. 1984. The complexity and stability of ecosystems. *Nature* 307:321–326.

Wharton, W. G., and K. H. Mann. 1981. Relationship between destructive grazing by the sea urchin, *Strongylocentrotus droebachiensis,* and the abundance of American lobster, *Homarus americanus,* on the Atlantic coast of Nova Scotia. *Canadian Journal of Fisheries and Aquatic Sciences* 38:1339–1349.

CHAPTER 8

Binns, W. O., and D. B. Redfern. 1983. Acid rain and forest decline in West Germany. *Forestry Commission Research and Development Paper* No. 131, Edinburgh.

Bormann, F. H., and G. E. Likens. 1979. *Pattern and Process in a Forested Ecosystem.* Springer-Verlag, New York.

Chapin, F. S., III. 1980. The mineral nutrition of wild plants. *Annual Review of Ecology and Systematics* 11:233–260.

Clymo, R. S. 1984. The limits to peat bog growth. *Philosophical Transactions of the Royal Society of London,* Series B, 303:605–654.

Etherington, J. R. 1975. *Environment and Plant Ecology.* Wiley, New York. Copyright 1975 by John Wiley & Sons, Ltd., London. Reprinted by permission.

Feller, M. C. 1980. Biomass and nutrient distribution in two eucalypt forest ecosystems. *Australian Journal of Zoology* 5:309–333.

Gorham, E., P. M. Vitousek, and W. A. Reiners. 1979. The regulation of chemical budgets over the course of terrestrial ecosystem succession. *Annual Review of Ecology and Systematics* 10:53–84.

Jordan, C. F., and R. Herrera. 1981. Tropical rain forests: are nutrients really critical? *American Naturalist* 117:167–180.

Likens, G. E., F. H. Bormann, and N. M. Johnson. 1981. Interactions between major biogeochemical cycles in terrestrial ecosystems. In *Some Perspectives of the Major Biogeochemical Cycles,* ed. G. E. Likens, pp. 93–123. Wiley, New York.

Likens, G. E., F. H. Bormann, N. M. Johnson, D. W. Fisher, and R. S. Pierce. 1970. Effects of forest cutting and herbicide treatment on nutrient budgets in the Hubbard Brook watershed-ecosystem. *Ecological Monographs* 40:23–47.

Likens, G. E., F. H. Bormann, R. S. Pierce, J. S. Eaton, and N. M. Johnson. 1977. *Biogeochemistry of a Forested Ecosystem.* Springer-Verlag, New York.

Likens, G. E., F. H. Bormann, R. S. Pierce, and W. A. Reiners. 1978. Recovery of a deforested ecosystem. *Science* 199:492–496. Copyright 1978 by the American Association for the Advancement of Science.

Likens, G. E., and T. J. Butler. 1981. Recent acidification of precipitation in North America. *Atmospheric Environment* 15, No. 7. Copyright 1981 by Pergamon Press. Reprinted by permission.

Likens, G. E., R. F. Wright, J. N. Galloway, and T. J. Butler. 1979. Acid rain. *Scientific American* 241(4):43–51.

Moore, J. J., P. Powding, and B. Healy. 1975. Glenamoy, Ireland. In "Structure and Function of Tundra Ecosystems," ed. T. Rosswell and O. W. Heal, *Ecological Bulletin* (Stockholm) 20:321–343.

Moore, P. D., and D. J. Bellamy. 1973. *Peatlands.* Elek Science, London.

Ovington, J. D. 1962. Quantitative ecology and the woodland ecosystem concept. *Advances in Ecological Research* 1:103–192.

Schindler, D. W., K. H. Mills, D. F. Malley, D. L. Findlay, J. A. Shearer, I. J. Davies, M. A. Turner, G. A. Linsey, and D. R. Cruikshank. 1985. Long-term ecosystem stress: the effects of years of experimental acidification on a small lake. *Science* 228:1395–1401.

Sjors, H. 1961. Surface patterns in boreal peatland. *Endeavour* 20(80):217–224.

Vitousek, P. 1982. Nutrient cycling and nutrient use efficiency. *American Naturalist* 119:553–572.

CHAPTER 9

Borchert, J. R. 1950. The climate of the central North American grassland. *Annals of the Association of American Geographers* 40:1–39.

Bryson, R. A., and T. J. Murray. 1977. *Climates of Hunger.* University of Wisconsin Press, Madison.

Covey, C. 1984. The earth's orbit and the ice ages. *Scientific American* 250(2):42–50.

Fritts, H. C. 1976. *Tree Rings and Climate.* Academic Press, New York.

Järvinen, O., and S. Ulfstrand. 1980. Species turnover of a continental bird fauna: northern Europe, 1850–1970. *Oecologia* 46:186–195.

LaMarche, V. C., Jr. 1974. Paleoclimatic inferences from long tree-ring records. *Science* 183:1043–1048. Copyright 1974 by the American Association for the Advancement of Science.

Lamb, H. F. 1985. Palynological evidence for postglacial change in the position of tree limit in Labrador. *Ecological Monographs* 55:241–258.

Moore, P. D., and J. A. Webb. 1978. *An Illustrated Guide to Pollen Analysis.* Hodder and Stoughton, London.

Nichols, H. 1976. Historical aspects of northern Canadian treeline. *Arctic* 29:38–47.

Ritchie, J. C., L. C. Cwynar, and R. W. Spear. 1983. Evidence from north-west Canada for an early Holocene Milankovitch thermal maximum. *Nature* 305:126–128. Copyright 1983 by Macmillan Journals Limited. Reprinted by permission.

Southward, A. J. 1980. The western English Channel: an inconstant ecosystem? *Nature* 285:361–366. Copyright 1980 by Macmillan Journals Limited. Reprinted by permission.

Southward, A. J., E. I. Butler, and L. Pennycuick. 1975. Recent cyclic changes in climate and in abundance of marine life. *Nature* 253:714–717. Copyright 1975 by Macmillan Journals Limited. Reprinted by permission.

Whitehead, D. R. 1981. Late-Pleistocene vegetational changes in northeastern North Carolina. *Ecological Monographs* 51:451–471.

CHAPTER 10

Andow, D. A. 1985. Plant diversification and insect populations in agroecosystems. In *Alternative Methods of Pest Control,* ed. D. Pimentel. Cornell University, Ithaca, N.Y.

Arthur, W. 1982. The evolutionary consequences of interspecific competition. *Advances in Ecological Research* 12:127–187.

Bakker, R. T. 1983. The deer flees, the wolf pursues: incongruencies in predator-prey coevolution. In *Coevolution,* ed. D. J. Futuyma and M. Slatkin, pp. 350–382. Sinauer, Sunderland, Mass.

Brower, L. P. 1969. Ecological chemistry. *Scientific American* 220(2):22–29.

Cannell, M. G. R. 1985. Autumn frost damage on young *Picea sitchensis* 1. Occurrence of autumn frosts in Scotland compared with Western North America. *Forestry* 58:131–143.

Cannell, M. G. R., L. J. Sheppard, R. I. Smith, and M. B. Murray. 1985. Autumn frost damage on young *Picea sitchensis* 2. Shoot frost hardening, and the probability of frost damage in Scotland. *Forestry* 58:145–166.

Clausen, J., D. D. Keck, and W. M. Hiesey. 1948. Experimental studies on the nature of species. III. Environmental responses of climatic races of *Achillea. Carnegie Institute Publication* No. 581, Washington, D.C.

Dawkins, R., and J. R. Krebs. 1979. Arms races between and within species. *Proceedings of the Royal Society of London,* Series B, 205:489–511.

Herrera, C. M. 1982. Defense of ripe fruit from pests: its significance in relation to plant-disperser interactions. *American Naturalist* 120:218–241.

Janzen, D. H. 1966. Coevolution of mutualism between ants and acacias in Central America. *Evolution* 20:249–275.

Krebs, C. J. 1985. *Ecology: The Experimental Analysis of Distribution and Abundance.* Harper & Row, New York.

Lack, D. 1947. *Darwin's Finches.* Cambridge University Press, Cambridge.

Mooney, H. A. (ed.). 1977. *Convergent Evolution in Chile and California: Mediterranean Climate Ecosystems.* Dowden, Hutchinson and Ross, Stroudsburg, Pa.

Mooney, H. A., and E. L. Dunn. 1970. Convergent evolution of Mediterranean-climate evergreen sclerophyll shrubs. *Evolution* 24:292–303.

Murdoch, W. W. 1975. Diversity, complexity, stability and pest control. *Journal of Applied Ecology* 12:795–807.

Orians, G. H., and R. T. Paine. 1983. Convergent evolution at the community level. In *Coevolution,* ed. D. J. Futuyma and M. Slatkin, pp. 431–458. Sinauer Associates, Sunderland, Mass.

Tomback, D. F. 1982. Dispersal of whitebark pine seeds by Clark's nutcracker: a mutualism hypothesis. *Journal of Animal Ecology* 51:451–467.

Turesson, G. 1930. The selective effect of climate upon the plant species. *Hereditas* 14:99–152.

Vane-Wright, R. I. 1980. Mimicry and its unknown ecological consequences. In *The Evolving Biosphere,* ed. P. H. Greenwood, pp. 157–168. Cambridge University Press, New York.

Wickler, W. 1968. *Mimicry in Plants and Animals.* Weidenfeld and Nicolson, London.

INDEX